CHICAGO PUBLIC LIBRARY
HAROLD WASHINGTON LIBRARY CENTER

R0006981474

REF
QC 94
.B92
cop. 1

FORM 125 M

BUSINESS/SCIENCE/TECHNOLOGY DIVISION

The Chicago Public Library

SEP 8 1977

Received_____ _____

CONVERSION FACTORS

CONVERSION FACTORS

Metric System—Inch/Pound System

**Ready source of 1333 conversion factors
with Inch-MM-Inch conversion tables**

FRANCIS BUCHEK

VANTAGE PRESS
New York   Washington   Atlanta   Hollywood

REF
QC
94
.B92
cop. 1

FIRST EDITION

All rights reserved, including the right of reproduction
in whole or part in any form.

Copyright © 1976 by Francis Buchek

Published by Vantage Press, Inc.
516 West 34th Street, New York, New York 10001

Printed in the United States of America

Standard Book Number 533-02184-7
Library of Congress Card Catalog Number 75-44560

SEP 9 1977

B/5/T
R

Foreword

A few years ago I was working on some engineering problem where, in a computation, a conversion factor was needed. After wasting time on searching through many conversion charts, tables, etc., finally I found what I was looking for.

This fact prompted me to work on a conversion pamphlet which will have all of the possible existing factors and could be found in one ready source. The effort paid off and the time-saving in solving many problems was tremendous.

This pamphlet should be helpful not only to all engineers, technicians, designers, scientists, and businessmen, but also to all students in colleges, universities, and down to high schools.

December 1974   Francis Buchek

Contents

Preface

8000 years ago, man began to seek bases for measurement. The oldest material standards existing in the world go back to 7000 B.C. where in Egypt man-made weights of the archaic "bequa" system, cylindrical stones, were used.

4000 B.C.—The cubit, what is now 18.24 inches, was standardized. The law also established the "Meridian Mile," equivalent to 4000 cubits or 1000 Egyptian fathoms.

A.D.

1790 —The Metric System (units of measurement) was originated in France.

1884 —England joined the metric convention, which resulted in establishing the relationship of the British units to metric units.

1889 —The United States joined the International Bureau of Weights and Measures, twenty-three years after Congress had authorized the use of the metric system in the United States.

1951 —Egypt and Albania went metric.
1954 —Israel, Jordan, and China (Taiwan) went metric.
1955 —Sudan went metric.
1958 —India went metric.
1959 —Japan went metric.

1971 —In August, Congress received a recommendation for the U. S. to go metric.

1972   —In the fall, the Senate approved a special bill to convert the U.S.A. to the metric system of measure.

????   —The U.S.A. went metric (in some respects since 1866, for over 100 years, we have been on the Metric System).

One of the biggest advantages of the metric system is the ability to decimalize all of the units.

Almost 90% of the world's population are using just seven base units to cover any problem of measurement: length—meter; mass—kilogram; time—second; electric current—ampere; thermodynamic temperature—Kelvin; luminous intensity—candela; and amount of—mole. Three of the units will solve 75% of all measurement.

The Metric System S.I. (System International d'Unites) was adopted in 1960 by General Conference of Weights and Measures.

From early ages scientists and mathematicians were always interested in units to communicate their discoveries to each other and the world at large. Today, in all realms of science, industry, and commerce, the seven base units are not sufficient and the purpose of this publication is to have a ready reference which covers all of the possible unit conversions.

FRANCIS BUCHEK

CONVERSION FACTORS

1

Conversion Factors

A

To convert	into	multiply by
Abamperes	Amperes	10
Abamperes	Statamperes	3×10^{10}
Abampere/sq. cm.	Amperes/sq. inch	64.52
Abampere-turns	Ampere-turns	10
Abampere-turns	Gilberts	12.57
Abampere-turns /cm.	Abampere-turns/inch	25.40
Abcoulombs	Coulombs	10
Abcoulombs	Statcoulombs	2.998×10^{10}
Abcoulombs/sq. cm.	Coulombs/sq. inch	64.52
Abfarads	Farads	10^9
Abfarads	Microfarads	10^{15}
Abfarads	Statfarads	9×10^{20}
Abhenries	Henries	10^{-9}
Abhenries	Millihenries	10^{-6}
Abhenries	Stathenries	$1/9 \times 10^{-20}$
Abmhos/cubic cm.	Mhos/mil. foot	1.662×10^2
Abmhos/cubic cm.	Megmhos/cubic cm.	10^3
Abohms	Megohms	10^{-15}

1

To convert	into	multiply by
Abohms	Microhoms	10^{-3}
Abohms	Ohms	10^{-9}
Abohms	Statohms	$1/9 \times 10^{-20}$
Abohms/cubic cm.	Microhms/cubic cm.	10^{-3}
Abohms/cubic cm.	Ohms/mil. foot	6.015×10^{-3}
Abvolts	Statvolts	$1/3 \times 10^{-10}$
Abvolts	Volts	10^{-8}
Acre	Ares	40.469
Acre	Hectare or sq. hecto-meter	0.4047
Acre	Rods	160
Acre	Roods	4
Acre	Sq. cm.	4.047×10^{7}
Acre	Sq. chain (Gunters)	10
Acre	Sq. feet	43,560.0
Acre	Sq. in.	6.273×10^{6}
Acre	Sq. km.	4.047×10^{-3}
Acre	Sq. links (Gunters)	1×10^{5}
Acre	Sq. meters	4,046.87
Acre	Sq. miles	1.563×10^{-3}
Acre	Township	4.34×10^{-5}
Acres	Sq. varas	5,645.38
Acres	Sq. yards	4,840
Acre-feet	Cu. feet	43,560
Acre-feet	Gallons	3.259×10^{5}
Acre-feet	Cubic meters	1,233.49
Amperes	Abamperes	0.1
Amperes	Statamperes	3×10^{9}
Amperes/sq. cm.	Amperes/sq. inch	6.452
Amperes/sq. cm.	Amperes/sq. meter	10^{4}
Amperes/sq. inch	Amperes/sq. cm.	0.1550
Amperes/sq. inch	Statamperes/sq. cm.	4.650×10^{8}
Amperes/sq. inch	Amperes/sq. cm.	0.1550
Amperes/sq. inch	Amperes/sq. meter	1,550.0
Amperes/sq. meter	Amperes/sq. cm.	10^{-4}
Amperes/sq. meter	Amperes/sq. inch	6.452×10^{-4}
Ampere-hours (abs.)	Coulombs (abs.)	3,600.0
Ampere-hours	Faradays	0.03731

To convert	into	multiply by
Ampere-turns	Abampere-turns	0.1
Ampere-turns	Gilberts	1.257
Ampere-turns/cm.	Ampere-turns/in.	2.540
Ampere-turns/cm.	Ampere-turns/meter	100.0
Ampere-turns/cm.	Gilberts/cm.	1.257
Ampere-turns/in.	Abampere-turns/cm.	0.03937
Ampere-turns/in.	Ampere-turns/cm.	0.3937
Ampere-turns/in.	Ampere-turns/meter	39.37
Ampere-turns/in.	Gilberts/cm.	0.4950
Ampere-turns/meter	Ampere-turns/cm.	0.01
Ampere-turns/meter	Ampere-turns/in.	0.0254
Ampere-turns/meter	Gilbers/cm.	0.01257
Angstrom unit	Inch	3.939×10^{-9}
Angstrom unit	Meter	1×10^{-10}
Angstrom unit	Centimeter	1×10^{-8}
Angstrom unit	Micron or (Mu)	1×10^{-4}
Are	Acre (U.S.)	0.02471
Are	Sq. meters	100
Ares	Sq. yards	119.60.
Astronomical Unit	Kilometers	1.495×10^{8}
Atmosphere	Bar	1.0133×10^{6}
Atmosphere	Cm. of Mercury	76.0
Atmosphere	Dynes/sq. cm.	1.0133×10^{6}
Atmosphere	Ft. of Water (at 4°C)	33.899
Atmosphere	Grams/sq. cm.	1,033.3
Atmosphere	Inch of Mercury (at 0°C)	29.92
Atmosphere	Kg./sq. cm.	1.0333
Atmosphere	Kg./sq. meter	10,332
Atmosphere	Millimeters of Mercury (at 32°F)	760
Atmosphere	Oz./sq. ft.	3.386×10^{4}
Atmosphere	Pounds/sq. foot	2,116.22
Atmosphere	Pounds/sq. in.	14.696
Atmosphere	L. Ton/sq. ft.	0.945
Atmosphere	S. Ton/sq. ft.	1.0581
Atmosphere	Ton/sq. inch	0.007348
Atmosphere	Ft. of Water(at 60°F)	33.934

To convert	into	multiply by
Atmosphere	Ft. of Water (at 70°F)	33.967

B

Bags (cement)	Pounds (cement)	94
Barrels (U.S., dry)	Cu. inches	7,056
Barrels (U.S., dry)	Quarts (dry)	105.0
Barrels (U.S., liquid)	Gallons	31.5
Barrels (oil)	Gallons (oil)	42.0
Barrels (cement)	Bags	4
Barrels (U.S., liquid)	Cu. meters	0.11924
Barrels/day	Gallons/minute	0.02917
Bars	Atmosphere	9.869×10^{-7}
Bars	Dynes/sq. cm.	1
Bars	Kgs./sq. meter	1.020×10^{-2}
Bars	Pounds/sq. ft.	2.089×10^{-3}
Bars	Pounds/sq. inch	1.4504×10^{-5}
Board-feet	Cu. in.	144
Board-feet	Cu. ft.	1/12
Baryl	Dyne/sq. cm.	1.000
Boiler horsepower	BTU/hour	33,480
Boiler horsepower	Kilowatts	9.803
BTU	Liter-Atmosphere	10.409
BTU	Ergs	1.0550×10^{10}
BTU	Foot-lbs.	778.3
BTU	Gram-calories	252.0
BTU	Horsepower-hrs.	3.931×10^{-4}
BTU	Joules	1,054.8
BTU	Kilogram-calories	0.2520
BTU	Kilogram-meters	107.5
BTU	Kilowatt-hrs.	2.928×10^{-4}
BTU	Centigrade heat units	2.205
BTU	Pounds carbon to CO_2	6.88×10^{-5}

To convert	into	multiply by
BTU	Pounds water evaporated from and at 212°F	0.001036
BTU	Cu. foot-atmosphere	3,676
BTU/hr.	Foot-pounds/sec.	0.2162
BTU/hr.	Gram-cal./sec.	0.0700
BTU/hr.	Horsepower-hrs.	3.929×10^{-4}
BTU/hr.	Watts	0.2931
BTU/min.	Foot-lbs./sec.	12.96
BTU/min.	Horsepower	0.02356
BTU/min.	Kilowatts	0.01757
BTU/min.	Watts	17.57
BTU/sq. ft./min.	Watts/sq. in.	0.1221
BTU/lb./°F	Calories/Gram/°C	1
BTU/sq. ft./min.	Kilowatts/sq. foot	0.1758
BTU/sq. ft./sec. (for a temp. gradient of 1°F/inch)	Calories, gram (15°C)/sq. cm./sec. for a tem. gradient of 1°C/cm.	1.2405
BTU (60°F)/°F	Calories/°C	453.6
Bucket (Br., dry)	Cu. cm.	1.818×10^4
Bushel (U.S., dry)	Cu. ft.	1.2445
Bushels	Cu. in.	2,150.4
Bushels	Cu. meters	0.03524
Bushels	Liters	35.24
Bushels	Pecks	4.0
Bushels	Pints (dry)	64.0
Bushels	Quarts (dry)	32.0

C

Calories, gram (mean)	BTU (mean)	3.9685×10^{-3}
Calories, gram	Foot-pounds	3.087
Calories, gram	Joules	4.185
Calories, gram	Liter-atmosphere	4.130×10^{-2}
Calories, gram	Horsepower-hours	1.5591×10^{-6}

To convert	into	multiply by
Calories, kilogram	Kilowatts hours	0.0011626
Calories, kilogram /sec	Kilowatts	4.185
Candle, power (spherical)	Lumens	12.556
Candle/sq. cm.	Lamberts	3.142
Candle/sq. inch	Lamberts	0.4870
Carats (metric)	Grams	0.2
Centares (centiares)	Sq. meters	1.0
Centigrade	Fahrenheit	$(°C \times 9/5) + 32$
Centigrams	Grams	0.01
Centiliter	Ounce, fluid (U.S.)	0.3382
Centiliter	Cubic inch	0.6103
Centiliter	Drams	2.705
Centiliters	Liters	0.01
Centimeters	Angstrom units	1×10
Centimeters	Feet	0.03281
Centimeters	Inches	0.3937
Centimeters	Kilometers	10^{-5}
Centimeters	Meters	0.01
Centimeters	Microns	10,000
Centimeters	Miles	6.214×10^{-6}
Centimeters	Millimeters	10
Centimeters	Mils	393.7
Centimeters	Yards	1.094×10^{-2}
Centimeter-dynes	Cm.-grams	1.020×10^{-3}
Centimeter-dynes	Meters-kgs.	1.020×10^{-8}
Centimeter-dynes	Pound-feet	7.376×10^{-8}
Centimeter-grams	Cm.-dynes	980.7
Centimeter-grams	Meters-kgs.	10^{-5}
Centimeter-grams	Pound-feet	7.233×10^{-5}
Centimeters of Mercury	Atmospheres	0.013158
Centimeters of Mercury	Feet of Water (39.1°F)	0.4461
Centimeters of Mercury	Kg./sq. meter	136.0

To convert	into	multiply by
Centimeters of Mercury	Pounds/sq. feet	27.845
Centimeters of Mercury	Pounds/sq. inch	0.19337
Centimeters/sec.	Feet/min.	1.1968
Centimeters/sec.	Feet/sec.	0.03281
Centimeters/sec.	Kilometers/hour	0.036
Centimeters/sec.	Knots	0.01943
Centimeters/sec.	Meters/min.	0.6
Centimeters/sec.	Miles/hour	0.02237
Centimeters/sec.	Miles/min.	3.728×10^{-4}
Centimeters/sec. /sec.	Feet/sec./sec.	0.03281
Centimeters/sec. /sec.	Km./hr./sec.	0.036
Centimeters/sec. /sec.	Meters/sec./sec.	0.01
Centimeters/sec. /sec.	Miles/hr./sec.	0.02237
Chain	Inches	792.0
Chain	Meters	20.12
Chain (Surveyor or Gunter's)	Yards	22.00
Circular mils	Sq. cms.	5.067×10^{-6}
Circular mils	Sq. mils	0.7854
Circular mils	Sq. inches	7.854×10^{-7}
Circumference	Radians	6.283
Cords	Cubic feet	128
Cords	Cord feet	8
Cord feet	Cubic feet	16
Coulombs	Abcoulombs	0.1
Coulomb	Statcoulombs	2.998×10^{9}
Coulombs	Faradays	1.036×10^{-5}
Coulombs/sq. cm.	Coulombs/sq. inch	6.452
Coulombs/sq. cm.	Coulombs/sq. meter	10^{4}
Coulombs/sq. inch	Coulombs/sq. cm.	0.1550
Coulombs/sq. inch	Abcoulombs/sq. cm.	0.01550

To convert	into	multiply by
Coulombs/sq. inch	Statcoulombs /sq. cm.	4.650×10^8
Coulombs/sq. inch	Coulombs/sq. meter	1,550
Coulombs/sq. meter	Coulombs/sq. cm.	10^{-4}
Coulombs/sq. meter	Coulombs/sq. in.	6.452×10^{-4}
Cubic centimeters	Bushel	2.838×10^{-5}
Cubic centimeters	Dram	0.27053
Cubic centimeters	Cu. Feet	3.531×10^{-5}
Cubic centimeters	Cu. inches	0.06102
Cubic centimeters	Cu. meters	10^{-6}
Cubic centimeters	Cu. yards	1.308×10^{-6}
Cubic centimeters	Gallons (U.S., liquid)	2.642×10^{-4}
Cubic centimeters	Gallons (Brit.)	2.20×10^{-4}
Cubic centimeters	Gill	8.4535×10^{-3}
Cubic centimeters	Liters	0.001
Cubic centimeters	Gram of Water (at 4°C)	0.999973
Cubic centimeters	Ounces (U.S., liquid)	0.03381
Cubic centimeters	Peck	1.135×10^{-4}
Cubic centimeters	Pints (U.S., liquid)	2.115×10^{-3}
Cubic centimeters	Quarts (U.S., liquid)	1.057×10^{-3}
Cubic feet	Acre-foot	2.296×10^{-5}
Cubic feet	Board feet	12
Cubic feet	Barrels	0.238
Cubic feet	Barrels (U.S., dry)	0.245
Cubic feet	Bushels (dry)	0.8036
Cubic feet	Cord	7.81×10^{-3}
Cubic feet	Cord foot	0.0625
Cubic feet	Cu. cms.	28,320.0
Cubic feet	Cu. inches	1,728.0
Cubic feet	Cu. meters	0.02832
Cubic feet	Cu. yards	0.03704
Cubic feet	Gallons (U.S., liquid)	7.48052
Cubic feet	Gallons (Brit.)	6.229

To convert	into	multiply by
Cubic feet	Liters	28.3163
Cubic feet	Ounces (fluid)	957.51
Cubic feet	Pints (U.S., liquid)	59.844
Cubic feet	Pecks	3.214
Cubic feet	Pound dry air at 70°F	0.07495
Cubic feet	Pound dry air at 0°C	0.08074
Cubic feet	Pound saturated air at 70°F	0.07425
Cubic feet	Pound water at 4°C	62.425
Cubic feet	Pound water at 60°F	62.365
Cubic feet	Pound water at 70°F	62.300
Cubic feet	Pound water at 212°F	59.826
Cubic feet	Quarts (U.S., liquid)	29.922
Cubic feet	Ton, reg.	0.01
Cubic feet	Ton, ship. (U.S.)	0.025
Cubic feet	Ton, ship. (Brit.)	0.0238
Cubic feet/min.	Cu. cms./sec.	472.0
Cubic feet/min.	Gallons/sec.	0.1247
Cubic feet/min.	Liters/sec.	0.4720
Cubic feet/min.	Pounds of water /min.	62.43
Cubic feet/sec.	Million gals./day	0.646317
Cubic feet/sec.	Gallons/min.	448.831
Cubic feet-atmospheres	Foot-pounds	2,116.3
Cubic feet-atmospheres	Liter-atmospheres	28.316
Cubic feet of water	Pounds	62.37
Cubic inches	Cu. cms.	16.39
Cubic inches	Cu. feet	5.787×10^{-4}
Cubic inches	Cu. meters	1.639×10^{-5}
Cubic inches	Cu. yards	2.143×10^{-5}
Cubic inches	Gallons	4.329×10^{-3}
Cubic inches	Liters	0.01639

To convert	into	multiply by
Cubic inches	Mil-feet	1.061×10^5
Cubic inches	Pints (U.S., liquid)	0.03463
Cubic inches	Quarts (U.S., liquid)	0.01732
Cubic meters	Bushels (dry)	28.38
Cubic meters	Cu. cms.	10^6
Cubic meters	Cu. feet	35.31
Cubic meters	Cu. inches	61,023.0
Cubic meters	Cu. yards	1.308
Cubic meters	Gallons (U.S., liquid)	264.2
Cubic meters	Liters	1,000.0
Cubic meters	Pints (U.S., liquid)	2,113.0
Cubic meters	Quarts (U.S., liquid)	1,057.0
Cubic yards	Cu. cms.	7.646×10^5
Cubic yards	Cu. feet	27
Cubic yards	Cu. inches	46,656.0
Cubic yards	Cu. meters	0.7646
Cubic yards	Gallons	202.0
Cubic yards	Liters	764.6
Cubic yards	Pints (U.S., liquid)	1,615.9
Cubic yards	Quarts (U.S., liquid)	807.9
Cubic yards/min.	Cu. feet/sec.	0.45
Cubic yards/min.	Gallons/sec.	3.367
Cubic yards/min.	Liters/sec.	12.74
Curies	Disintegration /min.	2.2×10^{12}
Curies	Coulombs/min.	1.1×10^{12}

D

Dalton	Gram	1.650×10^{-24}
Days	Hours	24
Days	Minutes	1,440
Days	Seconds	86,400

To convert	into	multiply by
Decigrams	Grams	0.1
Deciliters	Liters	0.1
Decimeters	Meters	0.1
Degrees (angle)	Minutes	60.0
Degrees (angle)	Radians	0.01745
Degrees (angle)	Quadrants	0.01111
Degrees (angle)	Seconds	3,600.0
Degrees/sec.	Radians/sec.	0.01745
Degrees/sec.	Radians/min.	0.1667
Degrees/sec.	Revolutions /sec.	2.778×10^{-3}
Dekagrams	Grams	10.0
Dekaliters	Liters	10.0
Dekameters	Meters	10.0
Drams	Grams	1.7718
Drams	Grains	27.3437
Drams	Ounces	0.0625
Drams (apothecaries or Troy)	Ounces (avoirdupois)	0.1371429
Drams (apothecaries or Troy)	Ounces (Troy)	0.125
Drams (U.S. fluid or apoth.)	Cubic cm.	3.6967
Dyne/cm.	Erg/sq. millimeter	0.01
Dyne/sq. cm.	Atmospheres	9.869×10^{-7}
Dyne/sq. cm.	Inch of Mercury (at 0°C)	2.953×10^{-5}
Dyne/sq. cm.	Inch of Water (at 4°C)	4.015×10^{-4}
Dynes	Grams	1.020×10^{-3}
Dynes	Joules/cm.	10^{-7}
Dynes	Joules/meter (Newtons)	10^{-5}
Dynes	Kilograms	1.020×10^{-6}
Dynes	Poundals	7.233×10^{-5}
Dynes	Pounds	2.248×10^{-6}
Dynes/sq. cm.	Bars	1
Dynes	Kgm./sec.^{2}	0.0001

To convert	into	multiply by

E

Ell	Cm.	114.30
Ell	Inches	45
Em, Pica	Inches	0.167
Em, Pica	Cm.	0.4233
Erg/sec.	Dyne-cm./sec.	1.000
Ergs	BTU	9.480×10^{-11}
Ergs	Dyne-centimeters	1.0
Ergs	Foot-pounds	7.367×10^{-8}
Ergs	Gram-calories	0.2389×10^{-7}
Ergs	Gram-cms.	1.020×10^{-3}
Ergs	Horsepower-hours	3.7250×10^{-14}
Ergs	Joules	10^{-7}
Ergs	Kg.-calories	2.389×10^{-11}
Ergs	Kg.-meters	1.020×10^{-8}
Ergs	Kilowatt-hours	0.2778×10^{-13}
Ergs	Watt-hours	0.2778×10^{-10}
Ergs/sec.	BTU/min.	$5,688 \times 10^{-9}$
Ergs/sec.	Ft.-lbs./min.	4.427×10^{-6}
Ergs/sec.	Ft.-lbs./sec.	7.3756×10^{-8}
Ergs/sec.	Horsepower	1.341×10^{-10}
Ergs/sec.	Kg.-calories/min.	1.433×10^{-9}
Egrs/sec.	Kilowatts	10^{-10}

F

Farads	Abfarads	10^{-9}
Farads	Microfarads	10^{6}
Farads	Statfarads	9×10^{11}
Faraday/sec.	Ampere (absolute)	9.6500×10^{4}
Faradays	Ampere-hours	26.80
Faradays	Coulombs	9.649×10^{4}
Fathom	Meter	1.828804
Fathoms	Feet	6.0
Feet	Cable	1.39×10^{-3}

To convert	into	multiply by
Feet	Centimeters	30.480
Feet	Chain (Eng.)	0.010
Feet	Chain (Surv.)	0.0152
Feet	Fathom	0.167
Feet	Furlong	1.515×10^{-3}
Feet	Hands	3.0
Feet	Inches	12
Feet	Kilometers	3.048×10^{-4}
Feet	Link (Eng.)	1.0
Feet	Links (Surv.)	1.515
Feet	Meters·	0.3048
Feet	Microns	3.048×10^{5}
Feet	Miles (Naut.)	1.645×10^{-4}
Feet	Miles (Stat.)	1.894×10^{-4}
Feet	Millimeters	304.8
Feet	Mils	1.2×10^{4}
Feet	Rod	0.0606
Feet	Pace	0.40
Feet	Varas	0.36
Feet	Yards	0.333
Feet of Water	Atmospheres	0.02950
Feet of Water	Inch of Mercury	0.8826
Feet of Water	Kgs./sq. cm.	0.03048
Feet of Water	Kgs./sq. meter	304.8
Feet of Water	Pounds/sq. ft.	62.43
Feet of Water	Pounds/sq. inch	0.4335
Feet/min.	Cms./sec.	0.5080
Feet/min.	Feet/sec.	0.01667
Feet/min.	Feet/hour	60
Feet/min.	Inch/hour	720
Feet/min.	Inch/min.	12
Feet/min.	Inch/sec.	0.20
Feet/min.	Kms./hour	0.01829
Feet/min.	Knot	9.898×10^{-3}
Feet/min.	Kilometers/min.	3.048×10^{-4}
Feet/min.	Meters/min.	0.3048
Feet/min.	Miles/min.	1.894×10^{-4}
Feet/min.	Miles/hour	0.01136

To convert	into	multiply by
Feet/min.	Meters/sec.	5.080×10^{-3}
Feet/sec.	Cms./sec.	30.48
Feet/sec.	Kms./hour	1.097
Feet/sec.	Knots	0.5921
Feet/sec.	Meters/min.	18.29
Feet/sec.	Miles/hour	0.6818
Feet/sec.	Miles/min.	0.01136
Feet/sec./sec.	Cms./sec./sec.	30.48
Feet/sec./sec.	Kms./hr./sec.	1.097
Feet/sec./sec.	Meters/sec./sec.	0.3048
Feet/sec./sec.	Miles/hr./sec.	0.6818
Feet/100 feet	Percent grade	1.0
Foot-candle	Lumen/sq. meter	10.764
Foot-pounds	BTU	1.286×10^{-3}
Foot-pounds	Ergs	1.356×10^{7}
Foot-pounds	Foot-poundals	32.174
Foot-pounds	Gram-calories	0.3239
Foot-pounds	Hp.-hours	5.05×10^{-7}
Foot-pounds	Joules	1.356
Foot-pounds	Kg.-calories	3.24×10^{-4}
Foot-pounds	Kg.-meters	0.1383
Foot-pounds	Kilowatt-hours	3.766×10^{-7}
Foot-poundals	BTU	3.995×10^{-5}
Foot-poundals	Joules	0.04214
Foot-poundals	Liter-atmospheres	4.159×10^{-4}
Foot-pounds/min.	BTU/min.	1.286×10^{-3}
Foot-pounds/min.	Foot-pound/sec.	0.01667
Foot-pounds/min.	Horsepower	3.030×10^{-5}
Foot-pounds/min.	Kg. calories/min.	3.24×10^{-4}
Foot-pounds/min.	Kilowatts	2.260×10^{-5}
Foot-pounds/sec.	BTU/hr.	4.6263
Foot-pounds/sec.	BTU/min.	7.717×10^{-2}
Foot-pounds/sec.	Horsepower	1.818×10^{-3}
Foot-pounds/sec.	Kg. calories/min.	1.945×10^{-2}
Foot-pounds/sec.	Kilowatts	1.356×10^{-3}
Furlongs	Feet	660
Furlongs	Rods	40
Furlongs	Miles (U.S.)	0.125

G

Gallons	Acre-feet	3.069×10^{-6}
Gallons	Barrels (U.S., liquid)	0.03175
Gallons	Barrels (oil)	0.0238
Gallons	Bushel	0.1074
Gallons	Cu. cms.	3,785.43
Gallons	Cu. feet	0.13368
Gallons	Cu. inches	231
Gallons	Cu. meters	3.785×10^{-3}
Gallons	Cu. yards	4.951×10^{-3}
Gallons	Gallons (dry)	0.8594
Gallons	Gills	32
Gallons	Liters	3.7853
Gallons	Peck	0.430
Gallons	Ounces (U.S., liquid)	128
Gallons	Pints (liquid)	8
Gallons of Water	Pounds of Water (at 4°C)	8.345
Gallons of Water	Pounds of Water (at 60°F)	8.337
Gallons of Water	Pounds of Water (at 70°F)	8.32ᶝ
Gallons of Water	Pounds of Water (at 212°F)	7.999
Gallons	Quarts (liquid)	4
Gallons (Br., Imp. liquid)	Gallons (U.S., liquid)	1.20095
Gallons (U.S.)	Gallons (Imp.)	0.83267
Gallons/min.	Cu. feet/sec.	2.228×10^{-3}
Gallons/min.	Liters/sec.	0.06308
Gallons/min.	Cu. feet/hour	8.0208
Gallons/min.	Overflow rate (ft. /hr.) Area (sq. ft.)	8.0208
Gallons water/min.	Tons water/24 hours	6.0086
Gausses	Lines/sq. inch	6.452
Gausses	Webers/sq. cm.	10^{-8}

To convert	into	multiply by
Gausses	Webers/sq. inch	6.452×10^{-8}
Gausses	Webers/sq. meter	10^{-4}
Gilberts	Abampere-turns	0.07958
Gilberts	Ampere-turns	0.7958
Gilberts/cm.	Ampere-turns/cm.	0.7958
Gilberts/cm.	Ampere-turns/inch	2.021
Gilberts/cm.	Ampere-turns/meter	79.58
Gills (Brit.)	Cubic cm.	142.07
Gills	Liters	0.1183
Gills	Pints (liquid)	0.25
Grade	Radian	0.01571
Grains	Drams (avoirdupois)	0.03657143
Grains (Troy)	Grains (avdp.)	1.0
Grains (Troy)	Grams	0.06480
Grains (Troy)	Ounces (avdp.)	2.286×10^{-3}
Grains (Troy)	Pennyweight (Troy)	0.04167
Grains/U.S. gal.	Parts/million	17.118
Grains/U.S. gal.	Pounds/million gal.	142.86
Grains/Imp. gal.	Parts/million	14.286
Grams	Dynes	980.7
Grams	Grains (Troy)	15.43
Grams	Joules/cm.	9.807×10^{-5}
Grams	Joules/meter (Newtons)	9.807×10^{-3}
Grams	Kilograms	0.001
Grams	Milligrams	1,000
Grams	Ounces (avdp.)	0.03527
Grams	Ounces (Troy)	0.03215
Grams	Poundals	0.07093
Grams	Pounds	2.205×10^{-3}
Grams/cm.	Pounds/inch	5.600×10^{-3}
Grams/cu. cm.	Grains/cu. feet	4.37×10^{5}
Grams/cu. cm.	Grains/gal.	5.842×10^{4}
Grams/cu. cm.	Grains/cu. cm.	15.43
Grams/cu. cm.	Kg./cu. cm.	0.001
Grams/cu. cm.	Kg./cu. m., Gr./l.	1,000
Grams/cu. cm.	Kg./l., Gr./ml.	1.000027
Grams/cu. cm.	Ounces/gal.	133.52

To convert	into	multiply by
Grams/cu. cm.	Ounces/cu. feet	998.8
Grams/cu. cm.	Parts/million	1×10^6
Grams/cu. cm.	Pounds/cu. foot	62.425
Grams/cu. cm.	Pounds/cu. inch	0.03613
Grams/cu. cm.	Pounds/mil-foot	3.405×10^{-7}
Grams/cu. cm.	Pounds/gallon	8.345
Grams/cu. cm.	Pounds/gal. (dry)	9.711
Grams/cu. cm.	Pounds/gal. (Brit.)	10.022
Grams/cu. cm.	Pounds/cu. yard	1,686
Grams/cu. cm.	Tons (long) /cu. yard	0.7525
Crams/cu. cm.	Tons (short) /cu. yard	0.8428
Grams/cu. cm.	Tons (metric) /cu. m.	1.0
Grams/liter	Grains/gal.	58.417
Grams/liter	Pounds/1,000 gal.	8.345
Grams/liter	Pounds/cu. feet	0.062427
Grams/liter	Parts/million	1,000.0
Grams/sq. cm.	Pounds/sq. feet	2.0481
Gram-calories	BTU	3.9683×10^{-3}
Gram-calories	Ergs	4.1868×10^7
Gram-calories	Foot-pounds	3.0880
Gram-calories	Horsepower-hours	1.5596×10^{-6}
Gram-calories	Kilowatt-hours	1.1630×10^{-6}
Gram-calories	Watt-hours	1.1630×10^{-3}
Gram-calories/sec.	BTU/hours	14.286
Gram centimeters	BTU	9.297×10^{-8}
Gram centimeters	Ergs	980.7
Gram centimeters	Joules	9.807×10^{-5}
Gram centimeters	Kg.-cal.	2.343×10^{-8}
Gram centimeters	Kg. meters	10^{-5}

H

Hand	Cm.	10.16
Hectares	Acres	2.471

To convert	into	multiply by
Hectares	Sq. feet	1.076×10^5
Hectograms	Grams	100
Hectoliters	Liters	100
Hectometers	Meters	100
Hectowatts	Watts	100
Hemispheres (solid angle)	Sphere	0.5
Hemispheres (solid angle)	Spherical right angle	4
Hemispheres (solid angle)	Steradians	6.283
Henries	Abhenries	10^9
Henries	Millihenries	1,000
Henries	Stathenries	$1/9 \times 10^{-11}$
Hogsheads (British)	Cu. ft.	10.114
Hogsheads (U.S.)	Cu. ft.	8.42184
Hogsheads (U.S.)	Gallons (U.S.)	63
Horsepower	Horsepower-boiler	0.076
Horsepower	BTU/min.	42.44
Horsepower	Ergs/sec.	7.457×10^9
Horsepower	Foot-lbs./min.	33,000
Horsepower	Foot-lbs./sec.	550
Horsepower (metric) (542.5 ft. lb./sec.)	Horsepower (550 ft.-lb./sec.)	0.9863
Horsepower (550 ft. lb./sec.)	Horsepower (metric) (CV) (542.5 ft.-lb./sec.)	1.0139
Horsepower	Kg. calories/hour	641.3
Horsepower	Kg. calories/min.	10.68
Horsepower	Kg. meters/hour	2.740×10^5
Horsepower	Kilowatts	0.7457
Horsepower	Poncelet	0.761
Horsepower	Ton-refrig.	0.212
Horsepower	Watts	745.7
Horsepower (boiler)	BTU/hour	33.479

To convert	into	multiply by
Horsepower (boiler)	Kilowatts	9.803
Horsepower (Brit.)	BTU/min.	42.42
Horsepower (Brit.)	BTU/hour	2,545
Horsepower (Brit.)	Pounds carbon to CO_2/hr.	0.175
Horsepower (Brit.)	Pounds water evaporated/hour at 212°F	2.623
Horsepower-hrs.	BTU (mean)	2,545
Horsepower-hrs.	Ergs	2.6845×10^{13}
Horsepower-hrs.	Foot-lbs.	1.98×10^6
Horsepower-hrs.	Ft.-tons	990
Horsepower-hrs.	Cu. ft.-lb./sq. inch	1.375×10^4
Horsepower-hrs.	Gram-calories	641,190
Horsepower-hrs.	Joules	2.684×10^6
Horsepower-hrs.	Kg.-calories	641.1
Horsepower-hrs.	Kg.-meters	2.737×10^5
Horsepower-hrs.	Kilowatts-hrs.	0.7457
Horsepower-hrs.	Liter-atmospheres	2.649×10^4
Horsepower-hrs.	Watt-hrs. (abs.)	745.7
Horsepower-hrs.	Pounds C to CO_2	0.175
Hours	Minutes	60
Hours	Seconds	3,600
Hours	Days	4.167×10^{-2}
Hours	Weeks	5.952×10^{-3}
Hundredweights (long)	Pounds	112
Hundredweights (long)	Tons (long)	0.05
Hundredweights (short)	Ounces (avoirdupois)	1,600
Hundredweights (short)	Pounds	100
Hundredweights (short)	Tons (metric)	0.0453592
Hundredweights (short)	Tons (long)	0.0446429

I

Inches	Angstroms	2.54×10^8
Inches	Cable	1.157×10^{-4}
Inches	Centimeters	2.540
Inches	Chain (Eng.)	8.333×10^{-4}
Inches	Chain (Surv.)	1.263×10^{-3}
Inches	Fathom	0.0139
Inches	Foot, Link, (Eng.)	0.0833
Inches	Furlong	1.26×10^{-4}
Inches	Hand	0.25
Inches	Kilometer	2.540×10^{-5}
Inches	Link (Surv.)	0.126
Inches	Meters	2.540×10^{-2}
Inches	Miles (Naut.)	1.371×10^{-5}
Inches	Miles (Stat.)	1.578×10^{-5}
Inches	Millimeters	25.40
Inches	Mils	1,000.0
Inches	Microns	2.54×10^4
Inches	Pace	0.0333
Inches	Rod	5.05×10^{-3}
Inches	Span	0.111
Inches	Yards	2.778×10^{-2}
Inches	Varas	0.03
Inches of Mercury	Atmospheres	0.03342
Inches of Mercury	Feet of Water	1.133
Inches of Mercury	Kgs./sq. cm.	0.03453
Inches of Mercury	Kgs./sq. meter	345.3
Inches of Mercury	Pounds/sq. ft.	70.73
Inches of Mercury	Pounds/sq. inch	0.4912
Inches of Water (at 4°C)	Atmospheres	2.458×10^{-3}
Inches of Water (at 4°C)	Inches of Mercury	0.07355
Inches of Water (at 4°C)	Kgs./sq. cm.	2.540×10^{-3}

To convert	into	multiply by
Inches of Water (at 4°C)	Ounces/sq. inch	0.5781
Inches of Water (at 4°C)	Pounds/sq. ft.	5.204
Inches of Water (at 4°C)	Pounds/sq. inch	0.03613
International Ampere	Ampere (abs.)	0.9998
International Volt	Volts (abs.)	1.0003
International Volt	Joules (abs.)	1.593×10^{-19}
International Volt	Joules	9.654×10^4

J

Joules	BTU	9.480×10^{-4}
Joules	Ergs	10^7
Joules	Foot-pounds	0.7376
Joules	Kg.-calories	2.389×10^{-4}
Joules	Kg.-meters	0.1020
Joules	Watt-hours	2.778×10^{-4}
Joules (absolute)	Cu. foot-atmospheres	0.3485
Joules (absolute)	Kilowatt-hours	2.778×10^{-7}
Joules (absolute)	Liter-atmospheres	0.009869
Joules/cm.	Grams	1.020×10^4
Joules/cm.	Dynes	10^7
Joules/cm.	Joules/meter (Newtons)	100.0
Joules/cm.	Poundals	723.3
Joules/cm.	Pounds	22.48

K

Kilograms	Decagrams	100
Kilograms	Drams (Troy, Ap.)	257.2
Kilograms	Drams (Avoir.)	564.4

To convert	into	multiply by
Kilograms	Dynes	980,665
Kilograms	Grams	1,000
Kilograms	Joules/cm.	0.09807
Kilograms	Joules/meter (Newton)	9.807
Kilograms	Ounces (Troy, Ap.)	32.151
Kilograms	Ounces (Avoir.)	35.274
Kilograms	Pennywts.	643.0
Kilograms	Poundals	70.93
Kilograms	Pounds	2.2046
Kilograms	Pounds (Troy, Ap.)	2.6792
Kilograms	Scruples	771.6
Kilograms	Slug	0.0685
Kilograms	Stone (Brit.)	0.1575
Kilograms	Quintal (m.)	0.01
Kilograms	Cwt. (short)	0.022
Kilograms	Tons (long)	9.842×10^{-4}
Kilograms	Tons (metric)	1.0×10^{-3}
Kilograms	Tons (short)	1.102×10^{-3}
Kilograms	Cu. cm. water (at 4°C)	1,000.0
Kilograms /cu. meter	Grams/cu. cm.	0.001
Kilograms /cu. meter	Pounds/cu. ft.	0.06243
Kilograms /cu. meter	Pounds/cu. inch	3.613×10^{-5}
Kilograms /cu. meter	Pounds/mil-foot	3.405×10^{-10}
Kilograms/meter	Pounds/ft.	0.6720
Kilograms/sq. cm.	Dynes	980,665
Kilograms/sq. cm.	Atmospheres	0.9678
Kilograms/sq. cm.	Feet of water	32.81
Kilograms/sq. cm.	Inches of mercury	28.96
Kilograms/sq. cm.	Pounds/sq. ft.	2,048
Kilograms/sq. cm.	Pounds/sq. inch	14.22
Kilograms /sq. meter	Atmospheres	9.678×10^{-5}

To convert	into	multiply by
Kilograms /sq. meter	Bars	98.07×10^{-6}
Kilograms /sq. meter	Feet of water	3.281×10^{-3}
Kilograms /sq. meter	Inches of mercury	2.896×10^{-3}
Kilograms /sq. meter	Pounds/sq. ft.	0.2048
Kilograms /sq. meter	Pounds/sq. inch	1.422×10^{-3}
Kilograms/sq. mm	Kgs/sq. meter	10^6
Kilogram-calories	BTU	3.968
Kilogram-calories	Foot-pounds	3,088
Kilogram-calories	Horsepower-hours	1.560×10^{-3}
Kilogram-calories	Joules	4,186
Kilogram-calories	Kg. meters	426.9
Kilogram-calories	Kilojoules	4.186
Kilogram-calories	Kilowatt-hours	1.163×10^{-3}
Kilogram-meters	BTU	9.294×10^{-3}
Kilogram-meters	Ergs	9.804×10^7
Kilogram-meters	Foot-pounds	7.233
Kilogram-meters	Joules	9.804
Kilogram-meters	Kg.-calories	2.342×10^{-3}
Kilogram-meters	Kilowatt-hours	2.723×10^{-6}
Kilolines	Maxwells	1,000.0
Kiloliters	Liters	1,000.0
Kilometers	Centimeters	10^5
Kilometers	Feet	3,281
Kilometers	Inches	3.937×10^4
Kilometers	Meters	1,000.0
Kilometers	Miles	0.6214
Kilometers	Millimeters	10^6
Kilometers	Yards	1,094
Kilometers/hr.	Cms./sec.	27.78
Kilometers/hr.	Feet/min.	54.68
Kilometers/hr.	Feet/sec.	0.9113
Kilometers/hr.	Knots	0.5396
Kilometers/hr.	Meters/min.	16.67

To convert	into	multiply by
Kilometers/hr.	Miles/hr.	0.6214
Kilometers/hr./sec.	Cms./sec./sec.	27.78
Kilometers/hr./sec.	Ft./sec./sec.	0.9113
Kilometers/hr./sec.	Meters/sec./sec.	0.2778
Kilometers/hr./sec.	Miles/hr./sec.	0.6214
Kilometers/min.	Km./hr.	60
Kilowatts	BTU/min.	56.92
Kilowatts	Foot-lbs./min.	4.426×10^4
Kilowatts	Foot-lbs./sec.	737.6
Kilowatts	Horsepower	1.341
Kilowatts	Kg.-calories/min.	14.34
Kilowatts	Watts	1,000.0
Kilowatts-hrs.	BTU	3,413
Kilowatts-hrs.	Ergs	3.600×10^{13}
Kilowatts-hrs.	Foot-lbs.	2.655×10^6
Kilowatts-hrs.	Gram-calories	859,850
Kilowatts-hrs.	Horsepower-hrs.	1.341
Kilowatts-hrs.	Joules	3.6×10^6
Kilowatts-hrs.	Kg.-calories	860.5
Kilowatts-hrs.	Kg.-meters	3.671×10^5
Kilowatts-hrs.	Pound of water evaporated from and at 212°F	3.53
Kilowatts-hrs.	Pound of water raised from 62° to 212°F	22.75
Knots	Feet/hr.	6,080
Knots	Kilometer/hr.	1.8532
Knots	Nautical miles/hr.	1.0
Knots	Statute miles/hr.	1.151
Knots	Yards/hr.	2,027
Knots	Feet/sec.	1.689

L

Lamberts	Candles/sq. inch	2.054
League	Miles (approx.)	3.0

To convert	into	multiply by
Light year	Miles	5.9×10^{12}
Light year	Kilometers	9.46091×10^{12}
Lines/sq. cm.	Gausses	1
Lines/sq. inch	Gausses	0.1550
Lines/sq. inch	Webers/sq. cm.	1.550×10^{-9}
Lines/sq. inch	Webers/sq. inch	10^{-8}
Lines/sq. inch	Webers/sq. meter	1.550×10^{-5}
Links (Engineer's)	Inches	12
Links (Surveyor's)	Inches	7.92
Liters	Bushels (U.S., dry)	0.02838
Liters	Cu. cm.	1,000.0
Liters	Cu. feet	0.03581
Liters	Cu. inches	61.02
Liters	Cu. meters	0.001
Liters	Cu. yards	1.308×10^{-3}
Liters	Gallons (Brit.)	0.220
Liters	Gallons (U.S., liquid)	0.2642
Liters	Gills	8.4537
Liters	Drams	270.518
Liters	Ounces (fluid)	33.815
Liters	Peck	0.1135
Liters	Pints (U.S., liquid)	2.1134
Liters	Quarts (U.S., liquid)	1.0567
Liters	Grams of Water (at 4°C)	1,000.0
Liters	Pounds of Water (at 4°C)	2.2046
Liters	Pounds of Water (at 60°F)	2.2025
Liters	Pounds of Water (at 70°F)	2.2002
Liters	Pounds of Water (at 212°F)	2.1129
Liter-atmospheres	Cu. foot-atmospheres	0.03532
Liter-atmospheres	Foot-pounds	74.74
Liters/min.	Cu. ft./sec.	5.886×10^{-4}

To convert	into	multiply by
Liters/min.	Gals./sec.	4.403×10^{-3}
Log$_{10}$ N	Log N or Ln N	2.303
Log$_e$ N or Ln N	Log$_{10}$ N	0.4343
Lumens/sq. ft.	Foot-candles	1.0
Lumen	Spherical candle-power	0.07958
Lumen	Watt	0.001496
Lumen/sq. ft.	Lumen/sq. meter	10.76
Lux	Foot-candle	0.0929

M

Maxwells	Kilolines	10^{-3}
Maxwells	Webers	10^{-8}
Megadyn	Dynes	10^6
Megadyn	Kgm./sq. sec.	10
Megadyn	Kg.	1.0197
Megalines	Maxwells	10^6
Megmhos/cu. cm.	Abmhos/cm. cubed	10^{-3}
Megmhos/cm. cubed	Megmhos/in. cubed	2.540
Megmhos/cm. cubed	Mhos/mil. foot	0.1662
Megmhos/inch cubed	Megmhos/cm. cubed	0.3937
Megohms	Microohms	10^{12}
Megohms	Ohms	10^6
Meters	Angstroms	10^{10}
Meters	Cable	4.557×10^{-3}
Meters	Centimeters	100
Meters	Chain (Eng.)	0.0328
Meters	Chain (Surv.)	0.0497
Meters	Fathom	0.547
Meters	Feet	3.28083
Meters	Furlong	4.971×10^{-3}
Meters	Inches	39.37
Meters	Kilometers	0.001
Meters	Links (Surv.)	4.971
Meters	Microns	10^6

To convert	into	multiply by
Meters	Miles (Naut.)	5.396×10^{-4}
Meters	Miles (Stat.)	6.214×10^{-4}
Meters	Millimeters	1,000.0
Meters	Mils	3.937×10^4
Meters	Rod	0.199
Meters	Yards	1.0936
Meters	Varas	1.179
Meters/min.	Cms./sec.	1.667
Meters/min.	Feet/min.	3.281
Meters/min.	Feet/sec.	0.05468
Meters/min.	Kms./hr.	0.06
Meters/min.	Knots	0.03238
Meters/min.	Miles/hr.	0.03728
Meters/sec.	Feet/min.	196.8
Meters/sec.	Feet/sec.	3.281
Meters/sec.	Kilometers/hr.	3.6
Meters/sec.	Kilometers/min.	0.06
Meters/sec.	Miles/hr.	2.237
Meters/sec.	Miles/min.	0.03728
Meters/sec./sec.	Cms./sec./sec.	100.0
Meters/sec./sec.	Ft./sec./sec.	3.281
Meters/sec./sec.	Km./hr./sec.	3.6
Meters/sec./sec.	Miles/hr./sec.	2.237
Meters-kilograms	Cm.-dynes	9.807×10^7
Meters-kilograms	Cm.-grams	10^5
Meters-kilograms	Pound-feet	7.233
Mhos (mil, foot)	Abmhos/cm.	6.015×10^{-3}
Mhos (mil, foot)	Micromhos/cm.	6.015
Mhos (mil, foot)	Micromhos/in.	15.28
Microfarads	Abfarads	10^{-15}
Microfarads	Farads	10^{-6}
Microfarads	Statfarads	9×10^5
Micrograms	Grams	10^{-6}
Microliters	Liters	10^{-6}
Microhms	Abohms	10^3
Microhms	Megohms	10^{-12}
Microhms	Ohms	10^{-6}
Microhms	Statohms	$1/9 \times 10^{-17}$

To convert	into	multiply by
Microhms /cm. cubed	Abohms/cm. cubed	10^8
Microhms /cm. cubed	Microhms/in. cubed	0.3937
Microhms /cm. cubed	Ohms/mil. foot	6.015
Microhms /inch cubed	Microhms /cm. cubed	2.540
Microns	Angstrom units	1×10^4
Microns	Centimeters	1×10^{-4}
Microns	Meters	10^{-6}
Miles (naut.)	Feet	6,080.27
Miles (naut.)	Kilometers	1.853
Miles (naut.)	Meters	1,853.0
Miles (naut.)	Miles (statute)	1.1516
Miles (naut.)	Yards	2,027.0
Miles (statute)	Cables	7.333
Miles (statute)	Centimeters	1.609×10^5
Miles (statute)	Chains (Eng.)	52.8
Miles (statute)	Chains (Surv.)	80
Miles (statute)	Fathoms	880
Miles (statute)	Feet	5,280
Miles (statute)	Furlongs	8.0
Miles (statute)	Inches	6.336×10^4
Miles (statute)	Kilometers	1.6094
Miles (statute)	League	0.333
Miles (statute)	Links (Surv.)	8,000
Miles (statute)	Light Year	1.691×10^{-13}
Miles (statute)	Meters	1,609.35
Miles (statute)	Miles (naut.)	0.8684
Miles (statute)	Millimeters	1.609×10^6
Miles (statute)	Myriameter	0.1609
Miles (statute)	Parsec	5.582×10^{-14}
Miles (statute)	Rods	320
Miles (statute)	Yards	1,760.0
Miles (statute)	Varas	1,900.8
Miles/hr.	Cms./sec.	44.704
Miles/hr.	Feet/min.	88

To convert	into	multiply by
Miles/hr.	Feet/sec.	1.4667
Miles/hr.	Feet/hr.	5,280
Miles/hr.	Inch/sec.	17.60
Miles/hr.	Inch/min.	1,056
Miles/hr.	Inch/hr.	6.336×10^4
Miles/hr.	Kms./hr.	1.6093
Miles/hr.	Kms./min.	0.02682
Miles/hr.	Knots	0.8684
Miles/hr.	Meters/sec.	0.4470
Miles/hr.	Meters/min.	26.823
Miles/hr.	Miles/min.	0.1667
Miles/hr./sec.	Cms./sec./sec.	44.70
Miles/hr./sec.	Feet/sec./sec.	1.467
Miles/hr./sec.	Kms./hr./sec.	1.6093
Miles/hr./sec.	Meters/sec./sec.	0.4470
Miles/min.	Cms./sec.	2,682
Miles/min.	Feet/sec.	88
Miles/min.	Kms./min.	1.6093
Miles/min.	Knots	52.10
Miles/min.	Miles/hr.	60
Mil-feet	Cu. inches	9.425×10^{-6}
Milliers	Kilograms	1,000.0
Millimicrons	Meters	1×10^{-9}
Milligrams	Grains	0.01543236
Milligrams	Grams	0.001
Milligrams/liter	Parts/million	1.0
Millihenries	Abhenries	10^6
Millihenries	Henries	0.001
Millihenries	Stathenries	$1/9 \times 10^{-14}$
Milliliters	Liters	0.001
Milliliters	Cu. cms.	1.000027
Millimeters	Centimeters	0.1
Millimeters	Feet	3.281×10^{-3}
Millimeters	Inches	0.03937
Millimeters	Kilometers	10^{-6}
Millimeters	Meters	0.001
Millimeters	Miles	6.214×10^{-7}
Millimeters	Mils	39.37

To convert	into	multiply by
Millimeters	Yards	1.094×10^{-3}
Million gal./day	Cu. ft./sec.	1.54723
Mils	Centimeters	2.540×10^{-3}
Mils	Feet	8.333×10^{-5}
Mils	Inches	0.001
Mils	Kilometers	2.540×10^{-8}
Mils	Yards	2.778×10^{-5}
Miner's inches	Cu. ft./min.	1.5
Minims (Brit)	Cubic cm.	0.059192
Minims (U.S., fluid)	Cubic cm.	0.061612
Minutes (angle)	Degrees	0.01667
Minutes (angle)	Quadrants	1.852×10^{-4}
Minutes (angle)	Radians	2.909×10^{-4}
Minutes (angle)	Seconds	60
Months	Days	30.42
Months	Hours	730
Months	Minutes	43,800
Months	Seconds	2.628×10^{6}
Myriagrams	Kilograms	10.0
Myriameters	Kilometers	10.0
Myriawatts	Kilowatts	10.0

N

Nepers	Decibels	8.686
Newton	Dynes	10^{5}
Newton	Kg./sq. sec.	1.0
Newton	Kg.	0.10197

O

Ohm (international)	Ohm (absolute)	1.0005
Ohms	Abohms	10^{9}
Ohms	Megohms	10^{-6}
Ohms	Microhms	10^{6}
Ohms	Statohms	$1/9 \times 10^{-11}$

To convert	into	multiply by
Ohms (mil. foot)	Abohms-cm.	166.2
Ohms (mil. foot)	Microhms-cm.	0.1662
Ohms (mil. foot)	Microhms-in.	0.06545
Ounces	Drams	16.0
Ounces	Grains	437.5
Ounces	Grams	28.349527
Ounces	Pounds	0.0625
Ounces	Ounces (Troy)	0.9115
Ounces (apothecaries)	Ounces (Troy)	1.0000
Ounces	Tons (long)	2.790×10^{-5}
Ounces	Tons (metric)	2.835×10^{-5}
Ounces (fluid)	Cu. inches	1.805
Ounces (fluid)	Liters	0.02957
Ounces (Troy)	Grains	480.0
Ounces (Troy)	Grams	31.103481
Ounces (Troy)	Ounces (avdp.)	1.09714
Ounces (Troy)	Pennyweights (Troy)	20.0
Ounces (Troy)	Pounds (Troy)	0.08333
Ounces/sq. inch	Dynes/sq. cm.	4,309
Ounces/sq. inch	Pounds/sq. inch	0.0625
Overflow rate (ft./hr.)	Gals./min.	$0.12468 \times$ Area (sq. ft.)
$\dfrac{1}{\text{Overflow rate (ft./hr.)}}$	Sq. ft./gal./min.	8.0208

P

Parsec	Miles	19×10^{12}
Parsec	Kilometers	3.084×10^{13}
Parts/million	Grains/U.S. gal.	0.0584
Parts/million	Grains/Imp. gal.	0.07016
Parts/million	Pounds/million gal.	8.345
Pecks (British)	Cu. inches	554.6
Pecks (British)	Liters	9.091901

To convert	into	multiply by
Pecks (U.S.)	Bushels	0.25
Pecks (U.S.)	Cu. inches	537.605
Pecks (U.S.)	Liters	8.809582
Pecks (U.S.)	Quarts (dry)	8.0
Pennyweights (Troy)	Grains	24.0
Pennyweights (Troy)	Ounces (Troy)	0.05
Pennyweights (Troy)	Grams	1.55517
Pennyweights (Troy)	Pounds (Troy)	4.1667×10^{-3}
Pints (dry)	Cu. inches	33.60
Pints (liq.)	Cu. cm.	473.2
Pints (liq.)	Cu. feet	0.01671
Pints (liq.)	Cu. inches	28.87
Pints (liq.)	Cu. meters	4.732×10^{-4}
Pints (liq.)	Cu. yards	6.189×10^{-4}
Pints (liq.)	Gallons	0.125
Pints (liq.)	Liters	0.4732
Pints (liq.)	Quarts (liq.)	0.5
Planck's quantum	Erg-second	6.624×10^{-27}
Poise	Gram/cm. sec.	1.00
Poundals	Dynes	13,826
Poundals	Grams	14.10
Poundals	Joules/cm.	1.383×10^{-3}
Poundals	Joules/meter (Newton)	0.1383
Poundals	Kilograms	0.01410
Poundals	Pounds	0.03108
Pounds (avoirdupois)	Ounces (Troy)	14.5833
Pounds	Cwt. (short)	0.01
Pounds	Cu. ft. dry air (0°C, 1 atm.)	12.385
Pounds	Cu. in. water (4°C)	27.687
Pounds	Drams	256
Pounds	Drams (Troy, Ap.)	116.67

To convert	into	multiply by
Pounds	Dynes	44.4823×10^4
Pounds	Grains	7,000
Pounds	Grams	453.5924
Pounds	Joules/cm.	0.04448
Pounds	Joules/meter (Newton)	4.448
Pounds	Kip	1.0×10^{-3}
Pounds	Kilograms	0.4536
Pounds	Ounces	16.0
Pounds	Ounces (Troy)	14.5833
Pounds	Pennyweights	291.67
Pounds	Poundals	32.174
Pounds	Pounds (Troy)	1.21528
Pounds	Slug	0.031
Pounds	Stone (Brit.)	0.07143
Pounds	Scruples	350
Pounds	Quintal (m.)	4.536×10^{-3}
Pounds	Tons (short)	0.0005
Pounds	Tons (long)	4.464×10^{-4}
Pounds	Tons (metric)	4.536×10^{-4}
Pounds (Troy)	Grains	5,760.0
Pounds (Troy)	Grams	373.24177
Pounds (Troy)	Ounces (avdp.)	13.1657
Pounds (Troy)	Ounces (Troy)	12.0
Pounds (Troy)	Pennyweights (Troy)	240.0
Pounds (Troy)	Pounds (avdp.)	0.822857
Pounds (Troy)	Tons (long)	3.6735×10^{-4}
Pounds (Troy)	Tons (metric)	3.7324×10^{-4}
Pounds (Troy)	Tons (short)	4.1143×10^{-4}
Pounds of water	Cu. feet	0.01602
Pounds of water	Cu. inches	27.68
Pounds of water	Gallons	0.1198
Pounds of water /min.	Cu. feet/sec.	2.669×10^{-4}
Pound feet	Dynes-cm.	1.356×10^7
Pound feet	Grams-cm.	13,825.0
Pound feet	Kg.-meters	0.1383
Pound feet squared	Kg. cm. squared	421.3

To convert	into	multiply by
Pound feet squared	Pounds in. squared	144
Pound inches	Kilograms-cm.	1.154
Pound in. squared	Kg. cm. squared	2.926
Pound in. squared	Pound feet squared	6.945×10^{-3}
Pounds/cu. ft.	Grams/cu. cm.	0.01602
Pounds/cu. ft.	Grains/cu. cm.	0.247
Pounds/cu. ft.	Grains/gal.	935.9
Pounds/cu. ft.	Grains/cu. ft.	7,000
Pounds/cu. ft.	Kgs./cu. meter	16.019
Pounds/cu. ft.	Ounces/gal.	2.140
Pounds/cu. ft.	Ounces/cu. ft.	16
Pounds/cu. ft.	Pounds/cu. in.	5.787×10^{-4}
Pounds/cu. ft.	Pounds/cu. yard	27
Pounds/cu. ft.	Pounds/gal.	0.1337
Pounds/cu. ft.	Pounds/gal. (dry)	0.1556
Pounds/cu. ft.	Pounds/gal. (Brit.)	0.161
Pounds/cu. ft.	Ounces/cu. in.	9.259×10^{-3}
Pounds/cu. ft.	Parts/million	1.602×10^{4}
Pounds/cu. ft.	Pounds/mil. foot	5.456×10^{-9}
Pounds/cu. ft.	Tons (long)/cu. ft.	4.464×10^{-4}
Pounds/cu. ft.	Tons (short)/cu. ft.	5.0×10^{-4}
Pounds/cu. ft.	Tons (long)/cu. yard	0.01205
Pounds/cu. ft.	Tons (short)/cu. yard	0.0135
Pounds/cu. in.	Gms./cu. cm.	27.68
Pounds/cu. in.	Kgs./cu. meter	2.768×10^{4}
Pounds/cu. in.	Pounds/cu. ft.	1,728
Pounds/cu. in.	Pounds/mil. foot	9.425×10^{-6}
Pounds/ft.	Kgs./meter	1.488
Pounds/in.	Grams/cm.	178.6
Pounds/mil. foot	Gms./cu. cm.	2.306×10^{6}
Pounds/sq. ft.	Atmospheres	4.725×10^{-4}
Pounds/sq. ft.	Feet of water	0.01602
Pounds/sq. ft.	Inches of mercury	0.01414
Pounds/sq. ft.	Kg./sq. centimeter	4.882×10^{-4}
Pounds/sq. ft.	Kg./sq. meter	4.882
Pounds/sq. in.	Atmospheres	0.068046
Pounds/sq. in.	Dynes/sq. cm.	6.895×10^{4}

To convert	into	multiply by
Pounds/sq. in.	Bar	0.06805
Pounds/sq. in.	Feet of water (at 4°C)	2.307
Pounds/sq. in.	Feet of water (at 60°F)	2.309
Pounds/sq. in.	Feet of water (at 70°F)	2.311
Pounds/sq. in.	Gr./sq. cm.	70.307
Pounds/sq. in.	Inches of mercury	2.036
Pounds/sq. in.	Kg./sq. centimeter	0.07031
Pounds/sq. in.	Kg./sq. meter	703.07
Pounds/sq. in.	Ounces/sq. in.	16
Pounds/sq. in.	Ounces/sq. ft.	2,304
Pounds/sq. in.	Pounds/sq. ft.	144.0
Pounds/sq. in.	Tons (short)/sq. in.	5.0×10^{-4}
Pounds/sq. in.	Tons (long)/sq. ft.	0.06429
Pounds/sq. in.	Tons (short)/sq. ft.	0.072
Pounds/sq. in.	Millimeters of mercury (at 0°C)	51.715

Q

Quadrants (angle)	Degrees	90.0
Quadrants (angle)	Minutes	5,400.0
Quadrants (angle)	Radians	1.571
Quadrants (angle)	Seconds	3.24×10^5
Quarts (dry)	Cu. inches	67.20
Quarts (liq.)	Cu. cms.	946.4
Quarts (liq.)	Cu. feet	0.03342
Quarts (liq.)	Cu. inches	57.75
Quarts (liq.)	Cu. meters	9.464×10^{-4}
Quarts (liq.)	Cu. yards	1.238×10^{-3}
Quarts (liq.)	Gallons	0.25
Quarts (liq.)	Liters	0.9463
Quintals	Pounds	100
Quires	Sheets	25

To convert	into	multiply by

R

To convert	into	multiply by
Radians	Degrees	57.30
Radians	Minutes	3,438
Radians	Quadrants	0.6366
Radians	Seconds	2.063×10^5
Reams	Sheets	500
Radians/sec.	Degrees/sec.	57.30
Radians/sec.	Revolutions /min.	9.549
Radians/sec.	Revolutions/sec.	0.1592
Radians/sec./sec.	Revs./min./min.	573.0
Radians/sec./sec.	Revs./min./sec.	9.549
Radians/sec./sec.	Revs./sec./sec.	0.1592
Revolutions	Degrees	360.0
Revolutions	Quadrants	4.0
Revolutions	Radians	6.283
Revolutions/min.	Degrees/sec.	6.0
Revolutions/min.	Radians/sec.	0.1047
Revolutions/min.	Revs./sec.	0.01667
Revolutions /min./min.	Radians/sec./sec.	1.745×10^{-3}
Revolutions /min./min.	Revs./min./sec.	0.01667
Revolutions /min./min.	Revs./sec./sec.	2.778×10^{-4}
Revolutions/sec.	Degrees/sec.	360.0
Revolutions/sec.	Radians/sec.	6.283
Revolutions/sec.	Revs./min.	60.0
Revolutions/sec./sec.	Radians/sec. /sec.	6.283
Revolutions/sec./sec.	Revs./min./min.	3,600.0
Revolutions/sec./sec.	Revs./min./sec.	60.0
Rod	Chain (Gunters)	0.25
Rod	Meters	5.029
Rod (Surveyors)	Yards	5.5
Rods	Feet	16.5

S

To convert	into	multiply by
Scruples	Grains	20.0
Seconds (angle)	Degrees	2.778×10^{-4}
Seconds (angle)	Minute	0.01667
Seconds (angle)	Quadrants	3.087×10^{-6}
Secouds (angle)	Radians	4.848×10^{-6}
Slugs	Gee pounds (pounds of mass)	1.0
Slug	Kilograms	14.594
Slug	Pounds	32.17
Sphere (solid angle)	Steradians	12.57
Spherical right angle	Hemispheres	0.25
Spherical right angle	Spheres	0.125
Spherical right angle	Steradians	1.571
Square centimeters	Circular mils	1.973×10^{5}
Square centimeters	Sq. feet	1.076×10^{-3}
Square centimeters	Sq. inches	0.1550
Square centimeters	Sq. meters	0.0001
Square centimeters	Sq. miles	3.861×10^{-11}
Square centimeters	Sq. millimeters	100.0
Square centimeters	Sq. yards	1.196×10^{-4}
Square cm. (cm. sqd.)	Sq. inches (inches sqd.)	0.02402
Square feet	Acres	2.296×10^{-5}
Square feet	Circular mils	1.834×10^{8}
Square feet	Sq. cms.	929.03
Square feet	Sq. chain (Eng.)	1.0×10^{-4}
Square feet	Sq. chain (Surv.)	2.296×10^{-4}
Square feet	Sq. inches	144.0
Square feet	Sq. km.	9.290×10^{-8}
Square feet	Sq. link (Eng.)	1.0
Square feet	Sq. links (Surv.)	2.296
Square feet	Sq. meters	0.09290

To convert	into	multiply by
Square feet	Sq. miles	3.587×10^{-8}
Square feet	Sq. millimeters	9.290×10^4
Square feet	Sq. mils	1.44×10^8
Square feet	Sq. rod	3.673×10^{-3}
Square feet	Sq. yards	0.1111
Square feet	Sq. varas	0.1296
Square feet (feet sqd.)	Sq. inches (inches sqd.)	2.07×10^4
Square inches	Acre	1.524×10^{-7}
Square inches	Circular mils	1.273×10^6
Square inches	Sq. cms.	6.4526
Square inches	Sq. chain (Eng.)	6.944×10^{-7}
Square inches	Sq. chain (Surv.)	1.594×10^{-6}
Square inches	Sq. feet	6.944×10^{-3}
Square inches	Sq. km.	6.452×10^{-10}
Square inches	Sq. link (Surv.)	0.0159
Square inches	Sq. meter	6.452×10^{-4}
Square inches	Sq. mile	2.491×10^{-10}
Square inches	Sq. millimeters	645.163
Square inches	Sq. mils	1.0×10^6
Square inches	Sq. rod	2.551×10^{-5}
Square inches	Sq. yards	7.716×10^{-4}
Square inches (inches sqd.)	Sq. cm. (cm. sqd.)	41.62
Square inches (inches sqd.)	Sq. ft. (Feet sqd.)	4.823×10^{-5}
Square kilometers	Acres	274.1
Square kilometers	Sq. cms.	10^{10}
Square kilometers	Sq. ft.	10.76×10^6
Square kilometers	Sq. inches	1.550×10^9
Square kilometers	Sq. meters	10^6
Square kilometers	Sq. miles	0.3861
Square kilometers	Sq. yards	1.196×10^6
Square meters	Acres	2.471×10^{-4}
Square meters	Sq. cms.	10^4
Square meters	Sq. feet	10.76
Square meters	Sq. inches	1,550
Square meters	Sq. miles	3.861×10^{-7}

To convert	into	multiply by
Square meters	Sq. millimeters	10^6
Square meters	Sq. yards	1.196
Square miles	Acres	640.0
Square miles	Sq. feet	27.88×10^6
Square miles	Sq. kms.	2.590
Square miles	Sq. meters	2.590×10^6
Square miles	Sq. yards	3.098×10^6
Square miles	Sq. varas	3,613,040.45
Square millimeters	Circular mils	1,973.0
Square millimeters	Sq. cms.	0.01
Square millimeters	Sq. feet	1.076×10^{-5}
Square millimeters	Sq. inches	1.550×10^{-3}
Square mils	Circular mils	1.273
Square mils	Sq. cms.	6.452×10^{-6}
Square mils	Sq. inches	10^{-6}
Square varas	Acres	0.0001771
Square varas	Sq. feet	7.716049
Square varas	Sq. miles	2.765×10^{-7}
Square varas	Sq. yards	0.857339
Square yards	Acres	2.066×10^{-4}
Square yards	Sq. cms.	8,361.0
Square yards	Sq. feet	9.0
Square yards	Sq. inches	1,296.0
Square yards	Sq. meters	0.8361
Square yards	Sq. miles	3.228×10^{-7}
Square yards	Sq. millimeters	8.361×10^5
Square yards	Sq. varas	1.1664
$\dfrac{1}{\text{Sq. ft./gal./min.}}$	Overflow rate (ft./hr.)	8.0208
Statamperes	Abamperes	$1/3 \times 10^{-10}$
Statamperes	Amperes	$1/3 \times 10^{-9}$
Statcoulombs	Abcoulombs	$1/3 \times 10^{-10}$
Statcoulombs	Coulombs	$1/3 \times 10^{-9}$
Statfarads	Abfarads	$1/9 \times 10^{-20}$
Statfarads	Farads	$1/9 \times 10^{-11}$
Statfarads	Microfarads	$1/9 \times 10^{-5}$
Stathenries	Abhenries	9×10^{20}

To convert	into	multiply by
Stathenries	Henries	9×10^{11}
Stathenries	Millihenries	9×10^{14}
Statohms	Abohms	9×10^{20}
Statohms	Megohms	9×10^{5}
Statohms	Microhms	9×10^{17}
Statohms	Ohms	9×10^{11}
Statvolts	Abvolts	3×10^{10}
Statvolts	Volts	300
Steradians	Hemispheres	0.1592
Steradians	Spheres	0.07958
Steradians	Spherical right angle	0.6366
Steres	Liters	10^{3}

T

To convert	into	multiply by
Temp. (°C) +273	Absolute temp. (°C)	1.0
Temp. (°C) +17.78	Temp. (°F)	1.8
Temp. (°F) +460	Absolute temp. (°F)	1.0
Temp. (°F)−32	Temp. (°C)	5/9
Tons (long)	Kilograms	1,016.0
Tons (long)	Pounds	2,240.0
Tons (long)	Tons (short)	1.120
Tons (metric)	Kilograms	1,000.0
Tons (metric)	Pounds	2,204.6
Tons (metric)	Tons (short)	1.1023
Tons (short)	Kilograms	907.1848
Tons (short)	Ounces	32,000.0
Tons (short)	Ounces (Troy)	29,166.66
Tons (short)	Pounds	2,000.0 (used in mechanics)
Tons (short)	Pounds (Troy)	2,430.56
Tons (short)	Tons (long)	0.89287
Tons (short)	Tons (metric)	0.90718

To convert	into	multiply by
Tons (British shipping)	Cu. ft.	42.00
Tons (U.S. shipping)	Cu. ft.	40.00
Tons refrigeration	BTU/hr.	12,000
Tons of water /24 hours	Pounds water/hour	83.333
Tons of water /24 hours	Gallon/min.	0.16643
Tons of water /24 hours	Cu. ft./hour	1.3349
Tons (short)/sq. ft.	Kg./sq. meter	9,765
Tons (short)/sq. ft.	Pounds/sq. inch	13.89
Tons (short)/sq. inch	Kg./sq. meter	1.406×10^6
Tons (short)/sq. inch	Pounds/sq. inch	2,000.0
Tons/sq. inch	Kg./sq. millimeter	1.5479

V

Varas	Foot	2.777
Varas	Inches	33.3333
Varas	Miles	5.26×10^{-4}
Varas	Yards	0.9259
Volts	Abvolts	10^8
Volts	Statvolts	0.003333
Volts (absolute)	Statvolts	0.003336
Volts/inch	Abvolts/cm.	3.937×10^7
Volts/inch	Volts/cm.	0.39370
Volts/inch	Statvolts/cm.	1.312×10^{-3}

W

Watts	BTU/hr.	3.4129
Watts	BTU/min.	0.05688

To convert	into	multiply by
Watts	Erg/sec.	10^7
Watts	Foot-lbs./min.	44.27
Watts	Foot-lbs./sec.	0.7376
Watts	Horsepower	1.341×10^{-3}
Watts	Horsepower (metric)	1.360×10^{-3}
Watts	Kg.-calories/min.	0.01434
Watts	Kilowatts	0.001
Watts (abs)	BTU (mean)/min.	0.056884
Watts (abs)	Joules/sec.	1
Watt-hours	BTU	3.413
Watt-hours	Ergs	3.60×10^{10}
Watt-hours	Foot-pounds	2,655.0
Watt-hours	Gram-calories	859.85
Watt-hours	Horsepower-hours	1.341×10^{-3}
Watt-hours	Kilogram-calories	0.8605
Watt-hours	Kilogram-meters	367.1
Watt-hours	Kilowatt-hours	0.001
Webers	Maxwells	10^8
Webers	Kilolines	10^5
Webers/sq. inch	Gausses	1.550×10^7
Webers/sq. inch	Lines/sq. inch	10^8
Webers/sq. inch	Webers/sq. cm.	0.1550
Webers/sq. inch	Webers/sq. meter	1,550.0
Webers/sq. meter	Gausses	10^4
Webers/sq. meter	Lines/sq. inch	6.452×10^4
Webers/sq. meter	Webers/sq. cm.	10^{-4}
Webers/sq. meter	Webers/sq. inch	6.452×10^{-4}
Weeks	Hours	168
Weeks	Minutes	10,080
Weeks	Seconds	604,800

Y

Yards	Centimeters	91.44
Yards	Feet	3
Yards	Inches	36
Yards	Kilometers	9.144×10^{-4}

To convert	into	multiply by
Yards	Meters	0.9144
Yards	Miles (naut.)	4.934×10^{-4}
Yards	Miles (stat.)	5.682×10^{-4}
Yards	Millimeters	914.4
Yards	Varas	1.08
Years (common)	Days	365
Years (common)	Hours	8,760
Years (leap)	Days	366
Years (leap)	Hours	8,784

2

Decimal Equivalents

of Inch and Conversion into Millimeters

Fraction	Decimal	Millimeter	Fraction	Decimal	Millimeter
1/64 ...	.015625	0,39688	33/64...	.515625	13,09690
1/32	.03125	0,79375	17/32	.53125	13,49378
3/64 ...	.046875	1,19063	35/64 ..	.546875	13,89065
1/16	.0625	1,58750	9/16	.5625	14,28753
5/64 ...	.078125	1,98438	37/64 ..	.578125	14,68440
3/32	.09375	2,38125	19/32	.59375	15,08128
7/64 ...	.109375	2,77813	39/64 ..	.609375	15,47816
1/8....................	.125	3,17501	5/8.....................	.625	15,87503
9/64 ...	.140625	3,57188	41/64 ..	.640625	16,27191
5/32	.15625	3,96876	21/32	.65625	16,66878
11/64 ...	.171875	4,36563	43/64 ..	.671875	17,06566
3/16	.1875	4,76251	11/16	.6875	17,46253
13/64 ...	.203125	5,15939	45/64 ..	.703125	17,85941
7/32	.21875	5,55626	23/32	.71875	18,25629
15/64 ...	.234375	5,95314	47/64 ..	.734375	18,65316
1/4.....................	.25	6,35001	3/4.....................	.75	19,05004
17/64 ...	.265625	6,74689	49/64 ..	.765625	19,44691
9/32	.28125	7,14376	25/32	.78125	19,84379
19/64 ...	.296875	7,54064	51/64 ..	.796875	20,24066
5/16	.3125	7,93752	13/16	.8125	20,63754
21/64 ...	.328125	8,33439	53/64 ..	.828125	21,03442
11/32	.34375	8,73127	27/32	.84375	21,43129
23/64 ...	.359375	9,12814	55/64 ..	.859375	21,82817
3/8.....................	.375	9,52502	7/8.....................	.875	22,22504
25/64 ...	.390625	9,92189	57/64 ..	.890625	22,62192
13/32	.40625	10,31877	29/32	.90625	23,01880
27/64 ...	.421875	10,71565	59/64 ..	.921875	23,41567
7/16	.4375	11,11252	15/16	.9375	23,81255
29/64 ...	.453125	11,50940	61/64 ..	.953125	24,20942
15/32	.46875	11,90627	31/32	.96875	24,60630
31/64 ...	.484375	12,30315	63/64 ..	.984375	25,00317
1/2.....................	.5	12,70002	1.....................	1.0	25,40005

NOTE:

1 inch (British) = 25,399 956 mm.
for industrial measurements, 1 inch (British) = 25,400 000 mm.

1 inch (American) = 25,400 051 mm.
for industrial measurements, 1 inch (American) = 25,400 000 mm.

3
Conversion of
Inches into Millimeters

Conversion of Inches into Millimeters

in.	0.0	0.1	0.2	0.3	0.4	0.5	0.6	0.7	0.8	0.9	in.
					Millimeters						
0		2.54	5.08	7.62	10.16	12.70	15.24	17.78	20.32	22.86	0
1	25.40	27.94	30.48	33.02	35.56	38.10	40.64	43.18	45.72	48.26	1
2	50.80	53.34	55.88	58.42	60.96	63.50	66.04	68.58	71.12	73.66	2
3	76.20	78.74	81.28	83.82	86.36	88.90	91.44	93.98	96.52	99.06	3
4	101.60	104.14	106.68	109.22	111.76	114.30	116.84	119.38	121.92	124.46	4
5	127.00	129.54	132.08	134.62	137.16	139.70	142.24	144.78	147.32	149.86	5
6	152.40	154.94	157.48	160.02	162.56	165.10	167.64	170.18	172.72	175.26	6
7	177.80	180.34	182.88	185.42	187.96	190.50	193.04	195.58	198.12	200.66	7
8	203.20	205.74	208.28	210.82	213.36	215.90	218.44	220.98	223.52	226.06	8
9	228.60	231.14	233.68	236.22	238.76	241.30	243.84	246.38	248.92	251.46	9
10	254.00	256.54	259.08	261.62	264.16	266.70	269.24	271.78	274.32	276.86	10

Basis 1 inch = 25,400 mm.

	Inches	0	1	2	3	4	5	6	7	8	9	10	11	12
			25.400	50.800	76.200	101.600	127.000	152.400	177.800	203.200	228.600	254.000	279.400	304.800
1/64	.015625	0.397	25.797	51.197	76.597	101.997	127.397	152.797	178.197	203.597	228.997	254.397	279.797	305.197
1/32	.01325	0.794	26.194	51.594	76.994	102.394	127.794	153.194	178.594	203.994	229.394	254.794	280.194	305.594
1/16	.0625	1.588	26.988	52.388	77.788	103.188	128.588	153.988	179.388	204.788	230.188	255.588	280.988	306.388
3/32	.09375	2.381	27.781	53.181	78.581	103.981	129.381	154.781	180.181	205.581	230.981	256.381	281.781	307.181
1/8	.125	3.175	28.575	53.975	79.375	104.775	130.175	155.575	180.975	206.375	231.775	257.175	282.575	307.975
5/32	.15625	3.969	29.369	54.769	80.169	105.569	130.969	156.369	181.769	207.169	232.569	257.969	283.369	308.769
3/16	.1875	4.763	30.163	55.563	80.963	106.363	131.763	157.163	182.563	207.963	233.363	258.763	284.163	309.563
7/32	.21875	5.556	30.956	56.356	81.756	107.156	132.556	157.956	183.356	208.756	234.156	259.556	284.956	310.356
1/4	.25	6.350	31.750	57.150	82.550	107.950	133.350	158.750	184.150	209.550	234.950	260.350	285.750	311.150
9/32	.28125	7.144	32.544	57.944	83.344	108.744	134.144	159.544	184.944	210.344	235.744	261.144	286.544	311.944
5/16	.3125	7.938	33.338	58.738	84.138	109.538	134.938	160.338	185.738	211.138	236.538	261.938	287.338	312.738
11/32	.34375	8.731	34.131	59.531	84.931	110.331	135.731	161.131	186.531	211.931	237.331	262.731	288.131	313.531
3/8	.375	9.525	34.925	60.325	85.725	111.125	136.525	161.925	187.325	212.725	238.125	263.525	288.925	314.325
13/32	.40625	10.319	35.719	61.119	86.519	111.919	137.319	162.719	188.119	213.519	238.919	264.319	289.719	315.119
7/16	.4375	11.113	36.513	61.913	87.313	112.713	138.113	163.513	188.913	214.313	239.713	265.113	290.513	315.913
15/32	.46875	11.906	37.306	62.706	88.106	113.506	138.906	164.306	189.706	215.106	240.506	265.906	291.306	316.706
1/2	.50	12.700	38.100	63.500	88.900	114.300	139.700	165.100	190.500	215.900	241.300	266.700	292.100	317.500
17/32	.53125	13.494	38.894	64.294	89.694	115.094	140.494	165.894	191.294	216.694	242.094	267.494	292.894	318.294
9/16	.5625	14.288	39.688	65.088	90.488	115.888	141.288	166.688	192.088	217.488	242.888	268.288	293.688	319.088
19/32	.59375	15.081	40.481	65.881	91.281	116.681	142.081	167.481	192.881	218.281	243.681	269.081	294.481	319.881
5/8	.625	15.875	41.275	66.675	92.075	117.475	142.875	168.275	193.675	219.075	244.475	269.875	295.275	320.675
21/32	.65625	16.669	42.069	67.469	92.869	118.269	143.669	169.069	194.469	219.869	245.269	270.669	296.069	321.469
11/16	.6875	17.463	42.863	68.263	93.663	119.063	144.463	169.863	195.263	220.663	246.063	271.463	296.863	322.263
23/32	.71875	18.256	43.656	69.056	94.456	119.856	145.256	170.656	196.056	221.456	246.856	272.256	297.656	323.056
3/4	.75	19.050	44.450	69.850	95.250	120.650	146.050	171.450	196.850	222.250	247.650	273.050	298.450	323.850
25/32	.78125	19.844	45.244	70.644	96.044	121.444	146.844	172.244	197.644	223.044	248.444	273.844	299.244	324.644
13/16	.8125	20.638	46.038	71.438	96.838	122.238	147.638	173.038	198.438	223.838	249.238	274.638	300.038	325.438
27/32	.84375	21.431	46.831	72.231	97.631	123.031	148.431	173.831	199.231	224.631	250.031	275.431	300.831	326.231
7/8	.875	22.225	47.625	73.025	98.425	123.825	149.225	174.625	200.025	225.425	250.825	276.225	301.625	327.025
29/32	.90625	23.019	48.419	73.819	99.219	124.619	150.019	175.419	200.819	226.219	251.619	277.019	302.419	327.819
15/16	.9375	23.813	49.213	74.613	100.013	125.413	150.813	176.213	201.613	227.013	252.413	277.813	303.213	328.613
31/32	.96875	24.606	50.006	75.406	100.806	126.206	151.606	177.006	202.406	227.806	253.206	278.606	304.006	329.406

Basis 1 inch = 25.400 mm.

Conversion of Inches into Millimeters

In.	0 0	1/16 .0625	1/8 .125	3/16 .1875	1/4 .25	5/16 .3125	3/8 .375	7/16 .4375	1/2 .5	9/16 .5625	5/8 .625	11/16 .6875	3/4 .75	13/16 .8125	7/8 .875	15/16 .9375
0	0.0	1.6	3.2	4.8	6.4	7.9	9.5	11.1	12.7	14.3	15.9	17.5	19.1	20.6	22.2	23.8
1	25.4	27.0	28.6	30.2	31.8	33.3	34.9	36.5	38.1	39.7	41.3	42.9	44.5	46.0	47.6	49.2
2	50.8	52.4	54.0	55.6	57.2	58.7	60.3	61.9	63.5	65.1	66.7	68.3	69.9	71.4	73.0	74.6
3	76.2	77.8	79.4	81.0	82.6	84.1	85.7	87.3	88.9	90.5	92.1	93.7	95.3	96.8	98.4	100.0
4	101.6	103.2	104.8	106.4	108.0	109.5	111.1	112.7	114.3	115.9	117.5	119.1	120.7	122.2	123.8	125.4
5	127.0	128.6	130.2	131.8	133.4	134.9	136.5	138.1	139.7	141.3	142.9	144.5	146.1	147.6	149.2	150.8
6	152.4	154.0	155.6	157.2	158.8	160.3	161.9	163.5	165.1	166.7	168.3	169.9	171.5	173.0	174.6	176.2
7	177.8	179.4	181.0	182.6	184.2	185.7	187.3	188.9	190.5	192.1	193.7	195.3	196.9	198.4	200.0	201.6
8	203.2	204.8	206.4	208.0	209.6	211.1	212.7	214.3	215.9	217.5	219.1	220.7	222.3	223.8	225.4	227.0
9	228.6	230.2	231.8	233.4	235.0	236.5	238.1	239.7	241.3	242.9	244.5	246.1	247.7	249.2	250.8	252.4
10	254.0	255.6	257.2	258.8	260.4	261.9	263.5	265.1	266.7	268.3	269.9	271.5	273.1	274.6	276.2	277.8
11	279.4	281.0	282.6	284.2	285.8	287.3	288.9	290.5	292.1	293.7	295.3	296.9	298.5	300.0	301.6	303.2
12	304.8	306.4	308.0	309.6	311.2	312.7	314.3	315.9	317.5	319.1	320.7	322.3	323.9	325.4	327.0	328.6
13	330.2	331.8	333.4	335.0	336.6	338.1	339.7	341.3	342.9	344.5	346.1	347.7	349.3	350.8	352.4	354.0
14	355.6	357.2	358.8	360.4	362.0	363.5	365.1	366.7	368.3	369.9	371.5	373.1	374.7	376.2	377.8	379.4
15	381.0	382.6	384.2	385.8	387.4	388.9	390.5	392.1	393.7	395.3	396.9	398.5	400.1	401.6	403.2	404.8
16	406.4	408.0	409.6	411.2	412.8	414.3	415.9	417.5	419.1	420.7	422.3	423.9	425.5	427.0	428.6	430.2
17	431.8	433.4	435.0	436.6	438.2	439.7	441.3	442.9	444.5	446.1	447.7	449.3	450.9	452.4	454.0	455.6
18	457.2	458.8	460.4	462.0	463.6	465.1	466.7	468.3	469.9	471.5	473.1	474.7	476.3	477.8	479.4	481.0
19	482.6	484.2	485.8	487.4	489.0	490.5	492.1	493.7	495.3	496.9	498.5	500.1	501.7	503.2	504.8	506.4

	0	1	2	3	4	5	6	7	8	9	10	11	12	13	14	15
21	533.4	535.0	536.6	538.2	539.8	541.3	542.9	544.5	546.1	547.7	549.3	550.9	552.5	554.0	555.6	557.2
22	558.8	560.4	562.0	563.6	565.2	566.7	568.3	569.9	571.5	573.1	574.7	576.3	577.9	579.4	581.0	582.6
23	584.2	585.8	587.4	589.0	590.6	592.1	593.7	595.3	596.9	598.5	600.1	601.7	603.3	604.8	606.4	608.0
24	609.6	611.2	612.8	614.4	616.0	617.5	619.1	620.7	622.3	623.9	625.5	627.1	628.7	630.2	631.8	633.4
25	635.0	636.6	638.2	639.8	641.4	642.9	644.5	646.1	647.7	649.3	650.9	652.5	654.1	655.6	657.2	658.8
26	660.4	662.0	663.6	665.2	666.8	668.3	669.9	671.5	673.1	674.7	676.3	677.9	679.5	681.0	682.6	684.2
27	685.8	687.4	689.0	690.6	692.2	693.7	695.3	696.9	698.5	700.1	701.7	703.3	704.9	706.4	708.0	709.6
28	711.2	712.8	714.4	716.0	717.6	719.1	720.7	722.3	723.9	725.5	727.1	728.7	730.3	731.8	733.4	735.0
29	736.6	738.2	739.8	741.4	743.0	744.5	746.1	747.7	749.3	750.9	752.5	754.1	755.7	757.2	758.8	760.4
30	762.0	763.6	765.2	766.8	768.4	769.9	771.5	773.1	774.7	776.3	777.9	779.5	781.1	782.6	784.2	785.8
31	787.4	789.0	790.6	792.2	793.8	795.3	796.9	798.5	800.1	801.7	803.3	804.9	806.5	808.0	809.6	811.2
32	812.8	814.4	816.0	817.6	819.2	820.7	822.3	823.9	825.5	827.1	828.7	830.3	831.9	833.4	835.0	836.6
33	838.2	839.8	841.4	843.0	844.6	846.1	847.7	849.3	850.9	852.5	854.1	855.7	857.3	858.8	860.4	862.0
34	863.6	865.2	866.8	868.4	870.0	871.5	873.1	874.7	876.3	877.9	879.5	881.1	882.7	884.2	885.8	887.4
35	889.0	890.6	892.2	893.8	895.4	896.9	898.5	900.1	901.7	903.3	904.9	906.5	908.1	909.6	911.2	912.8
36	914.4	916.0	917.6	919.2	920.8	922.3	923.9	925.5	927.1	928.7	930.3	931.9	933.5	935.0	936.6	938.2
37	939.8	941.4	943.0	944.6	946.2	947.7	949.3	950.9	952.5	954.1	955.7	957.3	958.9	960.4	962.0	963.6
38	965.2	966.8	968.4	970.0	971.6	973.1	974.7	976.3	977.9	979.5	981.1	982.7	984.3	985.8	987.4	989.0
39	990.6	992.2	993.8	995.4	997.0	998.5	1000.1	1001.7	1003.3	1004.9	1006.5	1008.1	1009.7	1011.2	1012.8	1014.4
40	1016.0	1017.6	1019.2	1020.8	1022.4	1023.9	1025.5	1027.1	1028.7	1030.3	1031.9	1033.5	1035.1	1036.6	1038.2	1039.8
41	1041.4	1043.0	1044.6	1046.2	1047.8	1049.3	1050.9	1052.5	1054.1	1055.7	1057.3	1058.9	1060.5	1062.0	1063.6	1065.2
42	1066.8	1068.4	1070.0	1071.6	1073.2	1074.7	1076.3	1077.9	1079.5	1081.1	1082.7	1084.3	1085.9	1087.4	1089.0	1090.6
43	1092.2	1093.8	1095.4	1097.0	1098.6	1100.1	1101.7	1103.3	1104.9	1106.5	1108.1	1109.7	1111.3	1112.8	1114.4	1116.0
44	1117.6	1119.2	1120.8	1122.4	1124.0	1125.5	1127.1	1128.7	1130.3	1131.9	1133.5	1135.1	1136.7	1138.2	1139.8	1141.4
45	1143.0	1144.6	1146.2	1147.8	1149.4	1150.9	1152.5	1154.1	1155.7	1157.3	1158.9	1160.5	1162.1	1163.6	1165.2	1166.8
46	1168.4	1170.0	1171.6	1173.2	1174.8	1176.3	1177.9	1179.5	1181.1	1182.7	1184.3	1185.9	1187.5	1189.0	1190.6	1192.2
47	1193.8	1195.4	1197.0	1198.6	1200.2	1201.7	1203.3	1204.9	1206.5	1208.1	1209.7	1211.3	1212.9	1214.4	1216.0	1217.6
48	1219.2	1220.8	1222.4	1224.0	1225.6	1227.1	1228.7	1230.3	1231.9	1233.5	1235.1	1236.7	1238.3	1239.8	1241.4	1243.0
49	1244.6	1246.2	1247.8	1249.4	1251.0	1252.5	1254.1	1255.7	1257.3	1258.9	1260.5	1262.1	1263.7	1265.2	1266.8	1268.4
50	1270.0	1271.6	1273.2	1274.8	1276.4	1277.9	1279.5	1281.1	1282.7	1284.3	1285.9	1287.5	1289.1	1290.6	1292.2	1293.8

4

Conversion of Thousandths of an Inch into Millimeters

Conversion of Thousandths of an Inch into Millimeters

Inches	0	1	2	3	4	5	6	7	8	9
	Millimeter									
.00	0	0.0254	0.0508	0.0762	0.1016	0.127	0.1524	0.1778	0.2032	0.2286
.01	0.254	0.2794	0.3048	0.3302	0.3556	0.381	0.4064	0.4318	0.4572	0.4826
.02	0.508	0.5334	0.5588	0.5842	0.6096	0.635	0.6604	0.6858	0.7112	0.7366
.03	0.762	0.7874	0.8128	0.8382	0.8636	0.889	0.9144	0.9398	0.9652	0.9906
.04	1.016	1.0414	1.0668	1.0922	1.1176	1.143	1.1684	1.1938	1.2192	1.2446
.05	1.27	1.2954	1.3208	1.3462	1.3716	1.397	1.4224	1.4478	1.4732	1.4986
.06	1.524	1.5494	1.5748	1.6002	1.6256	1.651	1.6764	1.7018	1.7272	1.7526
.07	1.778	1.8034	1.8288	1.8542	1.8796	1.905	1.9304	1.9558	1.9812	2.0066
.08	2.032	2.0574	2.0828	2.1082	2.1336	2.159	2.1844	2.2098	2.2352	2.2606
.09	2.286	2.3114	2.3368	2.3622	2.3876	2.413	2.4384	2.4638	2.4892	2.5146
.10	2.54	2.5654	2.5908	2.6162	2.6416	2.667	2.6924	2.7178	2.7432	2.7686
.11	2.794	2.8194	2.8448	2.8702	2.8956	2.921	2.9464	2.9718	2.9972	3.0226
.12	3.048	3.0734	3.0988	3.1242	3.1496	3.175	3.2004	3.2258	3.2512	3.2766
.13	3.302	3.3274	3.3528	3.3782	3.4036	3.429	3.4544	3.4798	3.5052	3.5306
.14	3.556	3.5814	3.6068	3.6322	3.6576	3.683	3.7084	3.7338	3.7592	3.7846
.15	3.81	3.8354	3.8608	3.8862	3.9116	3.937	3.9624	3.9878	4.0132	4.0386
.16	4.064	4.0894	4.1148	4.1402	4.1656	4.191	4.2164	4.2418	4.2672	4.2926
.17	4.318	4.3434	4.3688	4.3942	4.4196	4.445	4.4704	4.4958	4.5212	4.5466
.18	4.572	4.5974	4.6228	4.6482	4.6736	4.699	4.7244	4.7498	4.7752	4.8006
.19	4.826	4.8514	4.8768	4.9022	4.9276	4.953	4.9784	5.0038	5.0292	5.0546
.20	5.08	5.1054	5.1308	5.1562	5.1816	5.207	5.2324	5.2578	5.2832	5.3086
.21	5.334	5.3594	5.3848	5.4102	5.4356	5.461	5.4864	5.5118	5.5372	5.5626
.22	5.588	5.6134	5.6388	5.6642	5.6896	5.715	5.7404	5.7658	5.7912	5.8166
.23	5.842	5.8674	5.8928	5.9182	5.9436	5.969	5.9944	6.0198	6.0452	6.0706
.24	6.096	6.1214	6.1468	6.1722	6.1976	6.223	6.2484	6.2738	6.2992	6.3246
.25	6.35	6.3754	6.4008	6.4262	6.4516	6.477	6.5024	6.5278	6.5532	6.5786
.26	6.604	6.6294	6.6548	6.6802	6.7056	6.731	6.7564	6.7818	6.8072	6.8326
.27	6.858	6.8834	6.9088	6.9342	6.9596	6.985	7.0104	7.0358	7.0612	7.0866
.28	7.112	7.1374	7.1628	7.1882	7.2136	7.239	7.2644	7.2898	7.3152	7.3406
.29	7.366	7.3914	7.4168	7.4422	7.4676	7.493	7.5184	7.5438	7.5692	7.5946
.30	7.62	7.6454	7.6708	7.6962	7.7216	7.747	7.7724	7.7978	7.8232	7.8486
.31	7.874	7.8994	7.9248	7.9502	7.9756	8.001	8.0264	8.0518	8.0772	8.1026
.32	8.128	8.1534	8.1788	8.2042	8.2296	8.255	8.2804	8.3058	8.3312	8.3566
.33	8.382	8.4074	8.4328	8.4582	8.4836	8.509	8.5344	8.5598	8.5852	8.6106
.34	8.636	8.6614	8.6868	8.7122	8.7376	8.763	8.7884	8.8138	8.8392	8.8646
.35	8.89	8.9154	8.9408	8.9662	8.9916	9.017	9.0424	9.0678	9.0932	9.1186
.36	9.144	9.1694	9.1948	9.2202	9.2456	9.271	9.1964	9.3218	9.3472	9.3726
.37	9.398	9.4234	9.4488	9.4742	9.4996	9.525	9.5504	9.5758	9.6012	9.6266
.38	9.652	9.6774	9.7028	9.7282	9.7536	9.779	9.8044	9.8298	9.8552	9.8806
.39	9.906	9.9314	9.9568	9.9822	10.0076	10.033	10.0584	10.0838	10.1092	10.1346
.40	10.16	10.1854	10.2108	10.2362	10.2616	10.287	10.3124	10.3378	10.3632	10.3886
.41	10.414	10.4394	10.4648	10.4902	10.5156	10.541	10.5664	10.5918	10.6172	10.6426
.42	10.668	10.6934	10.7188	10.7442	10.7696	10.795	10.8204	10.8458	10.8712	10.8966
.43	10.922	10.9474	10.9728	10.9982	11.0236	11.049	11.0744	11.0998	11.1252	11.1506
.44	11.176	11.2014	11.2268	11.2522	11.2776	11.303	11.3284	11.3538	11.3792	11.4046
.45	11.43	11.4554	11.4808	11.5062	11.5316	11.557	11.5824	11.6078	11.6332	11.6586
.46	11.684	11.7094	11.7348	11.7602	11.7856	11.811	11.8364	11.8618	11.8872	11.9126
.47	11.938	11.9634	11.9888	12.0142	12.0396	12.065	12.0904	12.1158	12.1412	12.1666
.48	12.192	12.2174	12.2428	12.2682	12.2936	12.319	12.3444	12.3698	12.3952	12.4206
.49	12.446	12.4714	12.4968	12.5222	12.5476	12.573	12.5984	12.6238	12.6492	12.6746

NOTE: Use only values which can be read off directly. Compute intermediate values using factor 25.4.

Inches	0	1	2	3	4	5	6	7	8	9
					Millimeter					
.50	12.7	12.7254	12.7508	12.7762	12.8016	12.827	12.8524	12.8778	12.9032	12.9286
.51	12.954	12.9794	13.0048	13.0302	13.0556	13.081	13.1064	13.1318	13.1572	13.1826
.52	13.208	13.2334	13.2588	13.2842	13.3096	13.335	13.3604	13.3858	13.4112	13.4366
.53	13.462	13.4874	13.5128	13.5382	13.5636	13.589	13.6144	13.6398	13.6652	13.6906
.54	13.716	13.7414	13.7668	13.7922	13.8176	13.843	13.8684	13.8938	13.9192	13.9446
.55	13.97	13.9954	14.0208	14.0462	14.0716	14.097	14.1224	14.1478	14.1732	14.1986
.56	14.224	14.2494	14.2748	14.3002	14.3256	14.351	14.3764	14.4018	14.4272	14.4526
.57	14.478	14.5034	14.5288	14.5542	14.5796	14.605	14.6304	14.6558	14.6812	14.7066
.58	14.732	14.7574	14.7828	14.8082	14.8336	14.859	14.8844	14.9098	14.9352	14.9606
.59	14.986	15.0114	15.0368	15.0622	15.0876	15.113	15.1384	15.1638	15.1892	15.2146
.60	15.24	15.2654	15.2908	15.3162	15.3416	15.367	15.3924	15.4178	15.4432	15.4686
.61	15.494	15.5194	15.5448	15.5702	15.5956	15.621	15.6464	15.6718	15.6972	15.7226
.62	15.748	15.7734	15.7988	15.8242	15.8496	15.875	15.9004	15.9258	15.9512	15.9766
.63	16.002	16.0274	16.0528	16.0782	16.1036	16.129	16.1544	16.1798	16.2052	16.2306
.64	16.256	16.2814	16.3068	16.3322	16.3576	16.383	16.4084	16.4338	16.4592	16.4846
.65	16.51	16.5354	16.5608	16.5862	16.6116	16.637	16.6624	16.6878	16.7132	16.7386
.66	16.764	16.7894	16.8148	16.8402	16.8656	16.891	16.9164	16.9418	16.9672	16.9926
.67	17.018	17.0434	17.0688	17.0942	17.1196	17.145	17.1704	17.1958	17.2212	17.2466
.68	17.272	17.2974	17.3228	17.3482	17.3736	17.399	17.4244	17.4498	17.4752	17.5006
.69	17.526	17.5514	17.5768	17.6022	17.6276	17.653	17.6784	17.7038	17.7292	17.7546
.70	17.78	17.8054	17.8308	17.8562	17.8816	17.907	17.9324	17.9578	17.9832	18.0086
.71	18.034	18.0594	18.0848	18.1102	18.1356	18.161	18.1864	18.2118	18.2372	18.2626
.72	18.288	18.3134	18.3388	18.3642	18.3896	18.415	18.4404	18.4658	18.4912	18.5166
.73	18.542	18.5674	18.5928	18.6182	18.6436	18.669	18.6944	18.7198	18.7452	18.7706
.74	18.796	18.8214	18.8468	18.8722	18.8976	18.923	18.9484	18.9738	18.9992	19.0246
.75	19.05	19.0754	19.1008	19.1262	19.1516	19.177	19.2024	19.2278	19.2532	19.2786
.76	19.304	19.3294	19.3548	19.3802	19.4056	19.431	19.4564	19.4818	19.5072	19.5326
.77	19.558	19.5834	19.6088	19.6342	19.6596	19.685	19.7104	19.7358	19.7612	19.7866
.78	19.812	19.8374	19.8628	19.8882	19.9136	19.939	19.9644	19.9898	20.0152	20.0406
.79	20.066	20.0914	20.1168	20.1422	20.1676	20.193	20.2184	20.2438	20.2692	20.2946
.80	20.32	20.3454	20.3708	20.3962	20.4216	20.447	20.4724	20.4978	20.5232	20.5486
.81	20.574	20.5994	20.6248	20.6502	20.6756	20.701	20.7264	20.7518	20.7772	20.8026
.82	20.828	20.8534	20.8788	20.9042	20.9296	20.955	20.9804	21.0058	21.0312	21.0566
.83	21.082	21.1074	21.1328	21.1582	21.1836	21.209	21.2344	21.2598	21.2852	21.3106
.84	21.336	21.3614	21.3868	21.4122	21.4376	21.463	21.4884	21.5138	21.5392	21.5646
.85	21.59	21.6154	21.6408	21.6662	21.6916	21.717	21.7424	21.7678	21.7932	21.8186
.86	21.844	21.8694	21.8948	21.9202	21.9456	21.971	21.9964	22.0218	22.0472	22.0726
.87	22.098	22.1234	22.1488	22.1742	22.1996	22.225	22.2504	22.2758	22.3012	22.3266
.88	22.352	22.3774	22.4028	22.4282	22.4536	22.479	22.5044	22.5298	22.5552	22.5806
.89	22.606	22.6314	22.6568	22.6822	22.7076	22.733	22.7584	22.7838	22.8092	22.8346
.90	22.86	22.8854	22.9108	22.9362	22.9616	22.987	23.0124	23.0378	23.0632	23.0886
.91	23.114	23.1394	23.1648	23.1902	23.2156	23.241	23.2664	23.2918	23.3172	23.3426
.92	23.368	23.3934	23.4188	23.4442	23.4696	23.495	23.5204	23.5458	23.5712	23.5966
.93	23.622	23.6474	23.6728	23.6982	23.7236	23.749	23.7744	23.7998	23.8252	23.8506
.94	23.876	23.9014	23.9268	23.9522	23.9776	24.003	24.0284	24.0538	24.0792	24.1046
.95	24.13	24.1554	24.1808	24.2062	24.2316	24.257	24.2824	24.3078	24.3332	24.3586
.96	24.384	24.4094	24.4348	24.4602	24.4856	24.511	24.5364	24.5618	24.5872	24.6126
.97	24.638	24.6634	24.6888	24.7142	24.7396	24.765	24.7904	24.8158	24.8412	24.8666
.98	24.892	24.9174	24.9428	24.9682	24.9936	25.019	25.0444	25.0698	25.0952	25.1206
.99	25.146	25.1714	25.1968	25.2222	25.2476	25.273	25.2984	25.3238	25.3492	25.3746

5
Conversion of
Millimeters into Inches

Conversion of Millimeters into Inches

MM.	INCHES	INCHES	±
1	0.0394	1/32	+
2	0.0787	3/32	-
3	0.1181	1/8	-
4	0.1575	5/32	+
5	0.1969	3/16	+
6	0.2362	1/4	-
7	0.2756	9/32	-
8	0.3150	5/16	+
9	0.3543	11/32	+
10	0.3937	13/32	-
11	0.4331	7/16	-
12	0.4724	15/32	+
13	0.5118	1/2	+
14	0.5512	9/16	-
15	0.5906	19/32	-
16	0.6299	5/8	+
17	0.6693	21/32	+
18	0.7087	23/32	-
19	0.7480	3/4	-
20	0.7874	25/32	+
21	0.8268	13/16	+
22	0.8661	7/8	-
23	0.9055	29/32	-
24	0.9449	15/16	+

MM.	INCHES	INCHES	±
51	2.0079	2.0	+
52	2.0472	2-1/16	-
53	2.0866	2-3/32	-
54	2.1260	2-1/8	+
55	2.1654	2-5/32	+
56	2.2047	2-7/32	-
57	2.2441	2-1/4	-
58	2.2835	2-9/32	+
59	2.3228	2-5/16	+
60	2.3622	2-3/8	-
61	2.4016	2-13/32	-
62	2.4409	2-7/16	+
63	2.4803	2-15/32	+
64	2.5197	2-17/32	-
65	2.5591	2-9/16	-
66	2.5984	2-19/32	+
67	2.6378	2-5/8	+
68	2.6772	2-11/16	-
69	2.7165	2-23/32	-
70	2.7559	2-3/4	+
71	2.7953	2-25/32	+
72	2.8346	2-27/32	-
73	2.8740	2-7/8	-
74	2.9134	2-29/32	+

MM.	INCHES	INCHES	±
101	3.9764	3-31/32	+
102	4.0157	4-1/32	-
103	4.0551	4-1/16	-
104	4.0945	4-3/32	+
105	4.1339	4-1/8	+
106	4.1732	4-3/16	-
107	4.2126	4-7/32	-
108	4.2520	4-1/4	+
109	4.2913	4-9/32	+
110	4.3307	4-11/32	-
111	4.3701	4-3/8	-
112	4.4094	4-13/32	+
113	4.4488	4-7/16	+
114	4.4882	4-1/2	-
115	4.5276	4-17/32	-
116	3.5669	4-9/16	+
117	4.6063	4-19/32	+
118	4.6457	4-21/32	-
119	4.6850	4-11/16	-
120	4.7244	4-23/32	+
121	4.7638	4-3/4	+
122	4.8031	4-13/16	-
123	4.8425	4-27/32	-
124	4.8819	4-7/8	+

MM.	INCHES	INCHES	±
151	5.9449	5-15/16	+
152	5.9842	5-31/32	+
153	6.0236	6-1/32	-
154	6.0630	6-1/16	+
155	6.1024	6-3/32	-
156	6.1417	6-5/32	-
157	6.1811	6-3/16	-
158	6.2205	6-7/32	+
159	6.2598	6-1/4	+
160	6.2992	6-5/16	-
161	6.3386	6-11/32	-
162	6.3779	6-3/8	+
163	6.4173	6-13/32	+
164	6.4567	6-15/32	-
165	6.4961	6-1/2	-
166	6.5354	6-17/32	+
167	6.5748	6-9/16	+
168	6.6142	6-5/8	-
169	6.6535	6-21/32	-
170	6.6929	6-11/16	+
171	6.7323	6-23/32	+
172	6.7716	6-25/32	-
173	6.8110	6-13/16	-
174	6.8504	6-27/32	+

MM.	INCHES	INCHES	±
201	7.9134	7-29/32	+
202	7.9527	7-15/16	+
203	7.9921	8.0	-
204	8.0315	8-1/32	+
205	8.0709	8-1/16	+
206	8.1102	8-1/8	-
207	8.1496	8-5/32	-
208	8.1890	8-3/16	+
209	8.2283	8-7/32	+
210	8.2677	8-9/32	-
211	8.3071	8-5/16	-
212	8.3464	8-11/32	+
213	8.3858	8-3/8	+
214	8.4252	8-7/16	-
215	8.4646	8-15/32	-
216	8.5039	8-1/2	+
217	8.5433	8-17/32	+
218	8.5827	8-19/32	-
219	8.6220	8-5/8	-
220	8.6614	8-21/32	+
221	8.7008	8-11/16	+
222	8.7401	8-3/4	-
223	8.7795	8-25/32	-
224	8.8189	8-13/16	+

MM.	INCHES	INCHES	±
251	9.8819	9-7/8	+
252	9.9212	9-29/32	+
253	9.9606	9-31/32	-
254	10.0000	10.0	+
255	10.0393	10-1/32	-
256	10.0787	10-3/32	-
257	10.1181	10-1/8	-
258	10.1575	10-5/32	+
259	10.1968	10-3/16	+
260	10.2362	10-1/4	-
261	10.2756	10-9/32	-
262	10.3149	10-5/16	+
263	10.3543	10-11/32	+
264	10.3937	10-13/32	-
265	10.4330	10-7/16	-
266	10.4724	10-15/32	+
267	10.5118	10-1/2	+
268	10.5512	10-9/16	-
269	10.5905	10-19/32	-
270	10.6299	10-5/8	+
271	10.6693	10-21/32	+
272	10.7086	10-23/32	-
273	10.7480	10-3/4	-
274	10.7874	10-25/32	+

No.	Decimal	Fraction	±	No.	Decimal	Fraction	±	No.	Decimal	Fraction	±	No.	Decimal	Fraction	±	No.	Decimal	Fraction	±	No.	Decimal	Fraction	±
25	0.9843	31/32	+	75	2.9528	2-15/16	+	125	4.9213	4-29/32	+	175	6.8898	6-7/8	+	225	8.8583	8-27/32	+	275	10.8268	10-13/16	+
26	1.0236	1-1/32	-	76	2.9921	3.0	-	126	4.9606	4-31/32	-	176	6.9291	6-15/16	-	226	8.8976	8-29/32	-	276	10.8661	10-7/8	-
27	1.0630	1-1/16	+	77	3.0315	3-1/32	+	127	5.0000	5.0	+	177	6.9685	6-31/32	-	227	8.9370	8-15/16	-	277	10.9055	10-29/32	-
28	1.1024	1-3/32	+	78	3.0709	3-1/16	+	128	5.0394	5-1/32	+	178	7.0079	7.0	+	228	8.9764	8-31/32	+	278	10.9449	10-15/16	+
29	1.1417	1-5/32	-	79	3.1102	3-1/8	-	129	5.0787	5-3/32	-	179	7.0472	7-1/16	-	229	9.0157	9-1/32	-	279	10.9842	10-31/32	+
30	1.1811	1-3/16	-	80	3.1496	3-5/32	-	130	5.1181	5-1/8	-	180	7.0866	7-3/32	-	230	9.0551	9-1/16	-	280	11.0236	11-1/32	-
31	1.2205	1-7/32	+	81	3.1890	3-3/16	+	131	5.1575	5-5/32	+	181	7.1260	7-1/8	+	231	9.0945	9-3/32	+	281	11.0630	11-1/16	+
32	1.2598	1-1/4	+	82	3.2283	3-7/32	+	132	5.1968	5-3/16	+	182	7.1653	7-5/32	+	232	9.1338	9-1/8	+	282	11.1023	11-3/32	+
33	1.2992	1-5/16	-	83	3.2677	3-9/32	-	133	5.2362	5-1/4	-	183	7.2047	7-7/32	-	233	9.1732	9-3/16	-	283	11.1417	11-5/32	-
34	1.3386	1-11/32	-	84	3.3071	3-5/16	-	134	5.2756	5-9/32	-	184	7.2441	7-1/4	-	234	9.2126	9-7/32	-	284	11.1811	11-3/16	-
35	1.3780	1-3/8	+	85	3.3465	3-11/32	+	135	5.3150	5-5/16	+	185	7.2835	7-9/32	+	235	9.2520	9-1/4	+	285	11.2204	11-7/32	+
36	1.4173	1-13/32	+	86	3.3858	3-3/8	+	136	5.3543	5-11/32	+	186	7.3228	7-5/16	+	236	9.2913	9-9/32	+	286	11.2598	11-1/4	+
37	1.4567	1-15/32	-	87	3.4252	3-7/16	-	137	5.3937	5-13/32	-	187	7.3622	7-3/8	-	237	9.3307	9-11/32	-	287	11.2992	11-5/16	-
38	1.4961	1-1/2	-	88	3.4646	3-15/32	-	138	5.4331	5-7/16	-	188	7.4016	7-13/32	-	238	9.3701	9-3/8	-	288	11.3386	11-11/32	-
39	1.5354	1-17/32	+	89	3.5039	3-1/2	+	139	5.4724	5-15/32	+	189	7.4409	7-7/16	+	239	9.4094	9-13/32	+	289	11.3779	11-3/8	+
40	1.5748	1-9/16	+	90	3.5433	3-17/32	+	140	5.5118	5-1/2	+	190	7.4803	7-15/32	+	240	9.4488	9-7/16	+	290	11.4173	11-13/32	+
41	1.6142	1-5/8	-	91	3.5827	3-19/32	-	141	5.5512	5-9/16	-	191	7.5197	7-17/32	-	241	9.4882	9-1/2	-	291	11.4567	11-15/32	-
42	1.6535	1-21/32	-	92	3.6220	3-5/8	-	142	5.5905	5-19/32	-	192	7.5590	7-9/16	-	242	9.5275	9-17/32	-	292	11.4960	11-1/2	-
43	1.6929	1-11/16	+	93	3.6614	3-21/32	+	143	5.6299	5-5/8	+	193	7.5984	7-19/32	+	243	9.5669	9-9/16	+	293	11.5354	11-17/32	+
44	1.7323	1-23/32	+	94	3.7008	3-11/16	+	144	5.6693	5-21/32	+	194	7.6378	7-5/8	+	244	9.6063	9-19/32	+	294	11.5748	11-9/16	+
45	1.7717	1-25/32	-	95	3.7402	3-3/4	-	145	5.7087	5-23/32	-	195	7.6772	7-11/16	-	245	9.6457	9-21/32	-	295	11.6142	11-5/8	-
46	1.8110	1-13/16	-	96	3.7795	3-25/32	-	146	5.7480	5-3/4	-	196	7.7165	7-23/32	-	246	9.6850	9-11/16	-	296	11.6535	11-21/32	-
47	1.8504	1-27/32	+	97	3.8189	3-13/16	+	147	5.7874	5-25/32	+	197	7.7559	7-3/4	+	247	9.7244	9-23/32	+	297	11.6929	11-11/16	+
48	1.8898	1-7/8	+	98	3.8583	3-27/32	+	148	5.8268	5-13/16	+	198	7.7953	7-25/32	+	248	9.7638	9-3/4	+	298	11.7323	11-23/32	+
49	1.9291	1-15/16	-	99	3.8976	3-29/32	-	149	5.8661	5-7/8	-	199	7.8346	7-27/32	-	249	9.8031	9-13/16	-	299	11.7716	11-25/32	-
50	1.9685	1-31/32	-	100	3.9370	3-15/16	-	150	5.9055	5-29/32	-	200	7.8740	7-7/8	-	250	9.8425	9-27/32	-	300	11.8110	11-13/16	-

NOTE: The + or – sign indicates that the decimal equivalent is larger or smaller than the fractional equivalent.

59

6

Conversion of Millimeters into Inches (Decimals)

Conversion of Millimeters into Inches (Decimals)

Mm.	Inches	Mm.	Inches	Mm	Inches	Mm	Inches
1	.0394	46	1.8110	91	3.5827	136	5-3543
2	.0787	47	1.8504	92	3.6221	137	5.3937
3	.1181	48	1.8898	93	3.6614	138	5.4331
4	.1575	49	1.9291	94	3.7008	139	5.4725
5	.1969	50	1.9685	95	3.7402	140	5.5118
6	.2362	51	2.0079	96	3.7795	141	5.5512
7	.2756	52	2.0472	97	3.8189	142	5.5906
8	.3150	53	2.0866	98	3.8583	143	5.6299
9	.3543	54	2.1260	99	3.8976	144	5.6693
10	.3937	55	2.1654	100	3.9370	145	5.7087
11	.4331	56	2.2047	101	3.9764	146	5.7480
12	.4724	57	2.2441	102	4.0158	147	5.7874
13	.5118	58	2.2835	103	4.0551	148	5.8268
14	.5512	59	2.3228	104	4.0945	149	5.8662
15	.5906	60	2.3622	105	4.1339	150	5.9055
16	.6299	61	2.4016	106	4.1732	151	5.9449
17	.6693	62	2.4409	107	4.2126	152	5.9843
18	.7087	63	2.4803	108	4.2520	153	6.0236
19	.7480	64	2.5197	109	4.2913	154	6.0630
20	.7874	65	2.5591	110	4.3307	155	6.1024
21	.8268	66	2.5984	111	4.3701	156	6.1417
22	.8661	67	2.6378	112	4.4095	157	6.1811
23	.9055	68	2.6772	113	4.4488	158	6.2205
24	.9449	69	2.7165	114	4.4882	159	6.2599
25	.9843	70	2.7559	115	4.5276	160	6.2992
26	1.0236	71	2.7953	116	4.5669	161	6.3386
27	1.0630	72	2.8347	117	4.6063	162	6.3780
28	1.1024	73	2.8740	118	4.6457	163	6.4173
29	1.1417	74	2.9134	119	4.6850	164	6.4567
30	1.1811	75	2.9528	120	4.7244	165	6.4961
31	1.2205	76	2.9921	121	4.7638	166	6.5354
32	1.2598	77	3.0315	122	4.8032	167	6.5748
33	1.2992	78	3.0709	123	4.8425	168	6.6142
34	1.3386	79	3.1102	124	4.8819	169	6.6536
35	1.3780	80	3.1496	125	4.9213	170	6.6929
36	1.4173	81	3.1890	126	4.9606	171	6.7323
37	1.4567	82	3.2284	127	5.0000	172	6.7717
38	1.4961	83	3.2677	128	5.0394	173	6.8110
39	1.5354	84	3.3071	129	5.0787	174	6.8504
40	1.5748	85	3.3465	130	5.1181	175	6.8898
41	1.6142	86	3.3858	131	5.1575	176	6.9291
42	1.6535	87	3.4252	132	5.1969	177	6.9685
43	1.6929	88	3.4646	133	5.2362	178	7.0079
44	1.7323	89	3.5039	134	5.2756	179	7.0473
45	1.7717	90	3.5433	135	5.3150	180	7.0866

Mm.	Inches	Mm.	Inches	Mm.	Inches	Mm.	Inches
181	7.1260	226	8.8977	271	10.6693	316	12.4410
182	7.1654	227	8.9370	272	10.7087	317	12.4803
183	7.2047	228	8.9764	273	10.7481	318	12.5197
184	7.2441	229	9.0158	274	10.7874	319	12.5591
185	7.2835	230	9.0551	275	10.8268	320	12.5984
186	7.3228	231	9.0945	276	10.8662	321	12.6378
187	7.3622	232	9.1339	277	10.9055	322	12.6772
188	7.4016	233	9.1732	278	10.9449	323	12.7166
189	7.4410	234	9.2126	279	10.9843	324	12.7559
190	7.4803	235	9.2520	280	11.0236	325	12.7953
191	7.5197	236	9.2914	281	11.0630	326	12.8347
192	7.5591	237	9.3307	282	11.1024	327	12.8740
193	7.5984	238	9.3701	283	11.1418	328	12.9134
194	7.6378	239	9.4095	284	11.1811	329	12.9528
195	7.6772	240	9.4488	285	11.2205	330	12.9921
196	7.7165	241	9.4882	286	11.2599	331	13.0315
197	7.7559	242	9.5276	287	11.2992	332	13.0709
198	7.7953	243	9.5669	288	11.3386	333	13.1103
199	7.8347	244	9.6063	289	11.3780	334	13.1496
200	7.8740	245	9.6457	290	11.4173	335	13.1890
201	7.9134	246	9.6851	291	11.4567	336	13.2284
202	7.9528	247	9.7244	292	11.4961	337	13.2677
203	7.9921	248	9.7638	293	11.5355	338	13.3071
204	8.0315	249	9.8032	294	11.5748	339	13.3465
205	8.0709	250	9.8425	295	11.6142	340	13.3858
206	8.1103	251	9.8819	296	11.6536	341	13.4252
207	8.1496	252	9.9213	297	11.6929	342	13.4646
208	8.1890	253	9.9606	298	11.7323	343	13.5040
209	8.2284	254	10.0000	299	11.7717	344	13.5433
210	8.2677	255	10.0394	300	11.8110	345	13.5827
211	8.3071	256	10.0788	301	11.8504	346	13.6221
212	8.3465	257	10.1181	302	11.8898	347	13.6614
213	8.3858	258	10.1575	303	11.9292	348	13.7008
214	8.4252	259	10.1969	304	11.9685	349	13.7402
215	8.4646	260	10.2362	305	12.0079	350	13.7796
216	8.5040	261	10.2756	306	12.0473	351	13.8189
217	8.5433	262	10.3150	307	12.0866	352	13.8583
218	8.5827	263	10.3543	308	12.1260	353	13.8977
219	8.6221	264	10.3937	309	12.1654	354	13.9370
220	8.6614	265	10.4331	310	12.2047	355	13.9764
221	8.7008	266	10.4725	311	12.2441	356	14.0158
222	8.7402	267	10.5118	312	12.2835	357	14.0551
223	8.7795	268	10.5512	313	12.3229	358	14.0945
224	8.8189	269	10.5906	314	12.3622	359	14.1339
225	8.8583	270	10.6299	315	12.4016	360	14.1733

Mm.	Inches	Mm.	Inches	Mm.	Inches	Mm.	Inches
361	14.2126	406	15.9843	451	17.7559	496	19.5276
362	14.2520	407	16.0236	452	17.7953	497	19.5670
363	14.2914	408	16.0630	453	17.8347	498	19.6063
364	14.3307	409	16.1024	454	17.8740	499	19.6457
365	14.3701	410	16.1418	455	17.9134	500	19.6851
366	14.4095	411	16.1811	456	17.9528	501	19.7244
367	14.4488	412	16.2205	457	17.9922	502	19.7638
368	14.4882	413	16.2599	458	18.0315	503	19.8032
369	14.5276	414	16.2992	459	18.0709	504	19.8426
370	14.5670	415	16.3386	460	18.1103	505	19.8819
371	14.6063	416	16.3780	461	18.1496	506	19.9213
372	14.6457	417	16.4174	462	18.1890	507	19.9607
373	14.6851	418	16.4567	463	18.2284	508	20.0000
374	14.7244	419	16.4961	464	18.2677	509	20.0394
375	14.7638	420	16.5355	465	18.3071	510	20.0788
376	14.8032	421	16.5748	466	18.3465	511	20.1181
377	14.8425	422	16.6142	467	18.3859	512	20.1575
378	14.8819	423	16.6536	468	18.4252	513	20.1969
379	14.9213	424	16.6929	469	18.4646	514	20.2363
380	14.9607	425	16.7323	470	18.5040	515	20.2756
381	15.0000	426	16.7717	471	18.5433	516	20.3150
382	15.0394	427	16.8111	472	18.5827	517	20.3544
383	15.0788	428	16.8504	473	18.6221	518	20.3937
384	15.1181	429	16.8898	474	18.6614	519	20.4331
385	15.1575	430	16.9292	475	18.7008	520	20.4725
386	15.1969	431	16.9685	476	18.7402	521	20.5118
387	15.2362	432	17.0079	477	18.7796	522	20.5512
388	15.2756	433	17.0473	478	18.8189	523	20.5906
389	15.3150	434	17.0866	479	18.8583	524	20.6300
390	15.3544	435	17.1260	480	18.8977	525	20.6693
391	15.3937	436	17.1654	481	18.9370	526	20.7087
392	15.4331	437	17.2048	482	18.9764	527	20.7481
393	15.4725	438	17.2441	483	19.0158	528	20.7874
394	15.5118	439	17.2835	484	19.0552	529	20.8268
395	15.5512	440	17.3229	485	19.0945	530	20.8662
396	15.5906	441	17.3622	486	19.1339	531	20.9055
397	15.6299	442	17.4016	487	19.1733	532	20.9449
398	15.6693	443	17.4410	488	19.2126	533	20.9843
399	15.7087	444	17.4803	489	19.2520	534	21.0237
400	15.7481	445	17.5197	490	19.2914	535	21.0630
401	15.7874	446	17.5591	491	19.3307	536	21.1024
402	15.8268	447	17.5985	492	19.3701	537	21.1418
403	15.8662	448	17.6378	493	19.4095	538	21.1811
404	15.9055	449	17.6772	494	19.4489	539	21.2205
405	15.9449	450	17.7166	495	19.4882	540	21.2599

Mm.	Inches	Mm.	Inches	Mm.	Inches	Mm.	Inches
541	21.2992	586	23.0709	631	24.8426	676	26.6142
542	21.3386	587	23.1103	632	24.8819	677	26.6536
543	21.3780	588	23.1496	633	24.9213	678	26.6930
544	21.4174	589	23.1890	634	24.9607	679	26.7323
545	21.4567	590	23.2284	635	25.0000	680	26.7717
546	21.4961	591	23.2678	636	25.0394	681	26.8111
547	21.5355	592	23.3071	637	25.0788	682	26.8504
548	21.5748	593	23.3465	638	25.1182	683	26.8898
549	21.6142	594	23.3859	639	25.1575	684	26.9292
550	21.6536	595	23.4252	640	25.1969	685	26.9686
551	21.6930	596	23.4646	641	25.2363	686	27.0079
552	21.7323	597	23.5040	642	25.2756	687	27.0473
553	21.7717	598	23.5433	643	25.3150	688	27.0867
554	21.8111	599	23.5827	644	25.3544	689	27.1260
555	21.8504	600	23.6221	645	25.3937	690	27.1654
556	21.8898	601	23.6615	646	25.4331	691	27.2048
557	21.9292	602	23.7008	647	25.4725	692	27.2441
558	21.9685	603	23.7402	648	25.5119	693	27.2835
559	22.0079	604	23.7796	649	25.5512	694	27.3229
560	22.0473	605	23.8189	650	25.5906	695	27.3623
561	22.0867	606	23.8583	651	25.6300	696	27.4016
562	22.1260	607	23.8977	652	25.6693	697	27.4410
563	22.1654	608	23.9370	653	25.7087	698	27.4804
564	22.2048	609	23.9764	654	25.7481	699	27.5197
565	22.2441	610	24.0158	655	25.7874	700	27.5591
566	22.2835	611	24.0552	656	25.8268	701	27.5985
567	22.3229	612	24.0945	657	25.8662	702	27.6378
568	22.3622	613	24.1339	658	25.9056	703	27.6772
569	22.4016	614	24.1733	659	25.9449	704	27.7166
570	22.4410	615	24.2126	660	25.9843	705	27.7560
571	22.4804	616	24.2520	661	26.0237	706	27.7953
572	22.5197	617	24.2914	662	26.0630	707	27.8347
573	22.5591	618	24.3308	663	26.1024	708	27.8741
574	22.5985	619	24.3701	664	26.1418	709	27.9134
575	22.6378	620	24.4095	665	26.1811	710	27.9528
576	22.6772	621	24.4489	666	26.2205	711	27.9922
577	22.7166	622	24.4882	667	26.2599	712	28.0315
578	22.7559	623	24.5276	668	26.2993	713	28.0709
579	22.7953	624	24.5670	669	26.3386	714	28.1103
580	22.8347	625	24.6063	670	26.3780	715	28.1497
581	22.8741	626	24.6457	671	26.4174	716	28.1890
582	22.9134	627	24.6851	672	26.4567	717	28.2284
583	22.9528	628	24.7245	673	26.4961	718	28.2678
584	22.9922	629	24.7638	674	26.5355	719	28.3071
585	23.0315	630	24.8032	675	26.5748	720	28.3465

Mm.	Inches	Mm.	Inches	Mm.	Inches	Mm.	Inches
721	28.3859	766	30.1575	811	31.9292	856	33.7008
722	28.4252	767	30.1969	812	31.9686	857	33.7402
723	28.4646	768	30.2363	813	32.0079	858	33.7796
724	28.5040	769	30.2756	814	32.0473	859	33.8190
725	28.5434	770	30.3150	815	32.0867	860	33.8583
726	28.5827	771	30.3544	816	32.1260	861	33.8977
727	28.6221	772	30.3938	817	32.1654	862	33.9371
728	28.6615	773	30.4331	818	32.2048	863	33.9764
729	28.7008	774	30.4725	819	32.2442	864	34.0158
730	28.7402	775	30.5119	820	32.2835	865	34.0552
731	28.7796	776	30.5512	821	32.3229	866	34.0945
732	28.8189	777	30.5906	822	32.3623	867	34.1339
733	28.8583	778	30.6300	823	32.4016	868	34.1733
734	28.8977	779	30.6693	824	32.4410	869	34.2127
735	28.9371	780	30.7087	825	32.4804	870	34.2520
736	28.9764	781	30.7481	826	32.5197	871	34.2914
737	29.0158	782	30.7875	827	32.5591	872	34.3308
738	29.0552	783	30.8268	828	32.5985	873	34.3701
739	29.0945	784	30.8662	829	32.6379	874	34.4095
740	29.1339	785	30.9056	830	32.6772	875	34.4489
741	29.1733	786	30.9449	831	32.7166	876	34.4882
742	29.2126	787	30.9843	832	32.7560	877	34.5276
743	29.2520	788	31.0237	833	32.7953	878	34.5670
744	29.2914	789	31.0630	834	32.8347	879	34.6064
745	29.3308	790	31.1024	835	32.8741	880	34.6457
746	29.3701	791	31.1418	836	32.9134	881	34.6851
747	29.4095	792	31.1812	837	32.9528	882	34.7245
748	29.4489	793	31.2205	838	32.9922	883	34.7638
749	29.4882	794	31.2599	839	33.0316	884	34.8032
750	29.5276	795	31.2993	840	33.0709	885	34.8426
751	29.5670	796	31.3386	841	33.1103	886	34.8820
752	29.6064	797	31.3780	842	33.1497	887	34.9213
753	29.6457	798	31.4174	843	33.1890	888	34.9607
754	29.6851	799	31.4567	844	33.2284	889	35.0001
755	29.7245	800	31.4961	845	33.2678	890	35.0394
756	29.7638	801	31.5355	846	33.3071	891	35.0788
757	29.8032	802	31.5749	847	33.3465	892	35.1182
758	29.8426	803	31.6142	848	33.3859	893	35.1575
759	29.8819	804	31.6536	849	33.4253	894	35.1969
760	29.9213	805	31.6930	850	33.4646	895	35.2363
761	29.9607	806	31.7323	851	33.5040	896	35.2757
762	30.0001	807	31.7717	852	33.5434	897	35.3150
763	30.0394	808	31.8111	853	33.5827	898	35.3544
764	30.0788	809	31.8504	854	33.6221	899	35.3938
765	30.1182	810	31.8898	855	33.6615	900	35.4331

Mm.	Inches	Mm.	Inches	Mm.	Inches	Mm.	Inches
901	35.4725	926	36.4568	951	37.4410	976	38.4253
902	35.5119	927	36.4961	952	37.4804	977	38.4646
903	35.5512	928	36.5355	953	37.5198	978	38.5040
904	35.5906	929	36.5749	954	37.5591	979	38.5434
905	35.6300	930	36.6142	955	37.5985	980	38.5827
906	35.6694	931	36.6536	956	37.6379	981	38.6221
907	35.7087	932	36.6930	957	37.6772	982	38.6615
908	35.7481	933	36.7323	958	37.7166	983	38.7009
909	35.7875	934	36.7717	959	37.7560	984	38.7402
910	35.8268	935	36.8111	960	37.7953	985	38.7796
911	35.8662	936	36.8505	961	37.8347	986	38.8190
912	35.9056	937	36.8898	962	37.8741	987	38.8583
913	35.9449	938	36.9292	963	37.9135	988	38.8977
914	35.9843	939	36.9686	964	37.9528	989	38.9371
915	36.0237	940	37.0079	965	37.9922	990	38.9764
916	36.0631	941	37.0473	966	38.0316	991	39.0158
917	36.1024	942	37.0867	967	38.0709	992	39.0552
918	36.1418	943	37.1260	968	38.1103	993	39.0946
919	36.1812	944	37.1654	969	38.1497	994	39.1339
920	36.2205	945	37.2048	970	38.1890	995	39.1733
921	36.2599	946	37.2442	971	38.2284	996	39.2127
922	36.2993	947	37.2835	972	38.2678	997	39.2520
923	36.3386	948	37.3229	973	38.3072	998	39.2914
924	36.3780	949	37.3623	974	38.3465	999	39.3308
925	36.4174	950	37.4016	975	38.3859	1000	39.3701

Equivalents of Fractions of an Inch in Millimeters

mm. = inches x 25.399956

Inch.	Mm.	Inch.	Mm.	Inch.	Mm.	Inch.	Mm.
1/64	.3969	17/64	6.7469	33/64	13.0969	49/64	19.4468
1/32	.7937	9/32	7.1437	17/32	13.4937	25/32	19.8437
3/64	1.1906	10/64	7.5406	35/64	13.8906	51/64	20.2406
1/16	1.5875	5/16	7.9375	9/16	14.2875	13/16	20.6375
5/64	1.9844	21/64	8.3344	37/64	14.6843	53/64	21.0343
3/32	2.3812	11/32	8.7312	19/32	15.0812	27/32	21.4312
7/64	2.7781	23/64	9.1281	39/64	15.4781	55/64	21.8281
1/8	3.1750	3/8	9.5250	5/8	15.8750	7/8	22.2250
9/64	3.5719	25/64	9.9219	41/64	16.2718	57/64	22.6218
5/32	3.9687	13/32	10.3187	21/32	16.6687	29/32	23.0187
11/64	4.3656	27/64	10.7156	43/64	17.0656	59/64	23.4156
3/16	4.7625	7/16	11.1125	11/16	17.4625	15/16	23.8125
13/64	5.1594	29/64	11.5094	45/64	17.8593	61/64	24.2093
7/32	5.5562	15/32	11.9062	23/32	18.2562	31/32	24.6062
15/64	5.9531	31/64	12.3031	47/64	18.6531	63/64	25.0031
1/4	6.3500	1/2	12.7000	3/4	19.0500	1	25.4000